T0114628

David Leveson is an associate professor of Geology at Brooklyn College. Born in London in 1934, he was educated in England and the United States and received his Ph.D. from Columbia University in 1960. He has done field work in the western United States as well as in New York, Pennsylvania, Vermont, and the Channel Islands off the coast of France. As a result of his numerous field trips, Dr. Leveson has published several articles in professional journals and is currently working on a novel. His interest in photography, the results of which are partially represented in *A Sense of the Earth*, grew out of his involvement with the underground film movement. Dr. Leveson is married and lives in Brooklyn.

A Sense of the Earth

David Leveson

A Sense of the Earth

Photographs by the author
Drawings by Meg Leveson

Published for The American Museum of Natural History
The Natural History Press
Garden City, New York

1971

PUBLISHED BY DOUBLEDAY
a division of Random House, Inc.

A hardcover edition of this book was published in 1971 by Doubleday

The Natural History Press, publisher for The American Museum of Natural History, is a division of Doubleday & Company, Inc. Directed by a joint editorial board made up of members of the staff of both the Museum and Doubleday, the Natural History Press publishes books in all branches of the life and earth sciences, including anthropology and astronomy. The Natural History Press has its editorial offices at 277 Park Avenue, New York, New York 10017, and its business offices at 501 Franklin Avenue, Garden City, New York 11530.

www.doubleday.com

Book design by Earl Tidwell
Original jacket design by Douglas Wink
Original jacket photograph by David Leveson

The Library of Congress has cataloged the hardcover edition as:
Library of Congress Catalog Card Number 77-157608

Trade Paperback ISBN: 978-0-385-51489-7

146484122

For Ron Forbes who first suggested it, Ian Baldwin who convinced me I was serious, and Meg who helped idea become reality.

Foreword

Early man was imprinted by the soils, rocks, springs, and rivers among which he developed; his mind and bodily responses were shaped by the information he derived from his senses. Even if deficient in analytic, intellectual powers, he thus acquired unconsciously and inevitably a holistic, biologic kind of knowledge that enabled him to cope effectively with his natural environment—much as wild animals do in their native habitats. Very early in his evolution, furthermore, man developed another kind of knowledge not directly based on sensual experience. He searched for a form of reality different from the one he could touch, see, hear, smell, taste and otherwise apprehend directly with his senses. He imagined a Thing within the thing he actually experienced, a Force responsible for change and movement. He assumed that various components of the earth were endowed with an inner life that he symbolized by gods having activities and histories similar to his own.

These two different kinds of knowledge, one factual and the other conceptual, made him relate to nature in an intimate manner which was both organic and personalized.

Science has now introduced a third kind of knowledge of the

earth based on analytic description instead of holistic apprehension and symbolic interpretation. The new trend began with the systematic, conscious observation of the features of the land and with attempts to understand its history from the chemical composition and fossil contents of its layers. As time went on, geology became less and less concerned with the earth itself as directly experienced by the senses; it focused rather on the physics, chemistry, biology and evolution of the component parts of the earth. Modern geologists commonly deal with phenomena and properties that are not accessible to the unaided human senses and can be stated only in terms of abstract concepts meaningless to the non-specialist. An immense amount of precise knowledge has come from this extension of the senses by sophisticated equipment; but knowledge of specialized facts and laws *about* the earth cannot substitute for the satisfaction derived from knowing the earth organically, even when this abstract knowledge is synthesized intellectually. Modern geologists, like other modern scientists, run the risk of becoming alienated from what they seek to know.

David Leveson emphasizes in his book that it is essential and, fortunately, possible to recapture the totality of what he calls the geologic experience and to relate abstract geological knowledge to the direct human experience of the earth. The complete geologist should have "a foot firmly planted in each of two worlds: the modern scientific world, with all its abstractions and complex tools, and the world of the actual earth, with its primitive substances and concrete being." He should supplement analysis, specialization and abstraction with the worshipful attitude of the primitive man and with the approach of the nineteenth-century scholar who wandered through the hills, feeling the dirt and rocks under his feet and looking for visible clues to the history of our planet. All these varied expressions of the geologic experience are relevant to human life, not only because they provide useful technical information, but also, and perhaps even more, because they help man to understand his relation to other living things and especially to space and time.

One of the examples he discusses at length is the revolutionary change that has occurred in the concept of time since the middle of the nineteenth century under the influence of geologic discoveries.

In the past, man had little historical sense and lived as if time were an unending and undifferentiated present. Very early, in contrast, the geologist was forced to imagine an immense duration in order to allow for the profound change in land structure that obviously had occurred through extremely slow natural processes. Within absolute geological time—the span of years during which the earth has existed—geologists can define with greater and greater precision relative time, namely, the chronological order in which geologic events occurred. Absolute time stuns the imagination by its immensity—several billions of years! Relative time, on the other hand, introduces concepts that are even more dramatic in their implications; they relate man to other living forms that have disappeared and thus suggest that man also may disappear through natural processes, even if he does not destroy himself through misuse of his power.

Man could at first find comfort in the thought that all natural processes exhibit cyclic repetition; cycles embody the excitement of movement together with the security of returning to the same place or state. "To be part of a cycle is thus to be part of eternity, to hold a guarantee of immortality." But this concept too has been upset by geologic discoveries. The rock cycle, the water cycle and other natural cycles never repeat themselves exactly and are more like spirals or complex helixes. We too, like nature, can never go home again because when we try to retrace our steps we go up or down. This is perhaps another way of stating that everything in nature and in life is historical.

The geologist cannot escape dedication to history, and this makes him the epitome of Western man whose tragic sense of life comes from his awareness that he is not absolute and final but a part of nature that comes from somewhere and is going towards somewhere—but who knows where?

David Leveson states that the major reward of the geologist is "the deep intimacy that can arise between a man and his natural surroundings, an intimacy that stems from both detailed contact and continued theoretical and philosophic contemplation." The major reward from reading *A Sense of the Earth* is to discover that there is an underlying unity to the holistic knowledge and worshipful attitude of primitive man before nature, the total geologic experience of the nineteenth-century pioneers, and the abstruse analyses and attempts at abstract synthetic expressions of the modern scientists. Man's attempt to know and understand the earth has generated a theology of nature that incorporates both reverence and knowledge. Now that man knows he is literally an organic part of the earth, he understands why he should not exploit it selfishly but instead manage it with love and thus contribute to the grandeur of its evolution and of his own terrestrial experience.

René Dubos
The Rockefeller University
January, 1971

Contents

A Sense of the Earth

Introduction

At the brink of Canyon De Chelly I met a man searching for a place to live. He wanted a place, he said, connected to the earth, bound to the soil, to rock, where he could earn an honest living and begin to lead a meaningful life. As we chatted, his gaze kept shifting to the bottom of the canyon. There, separated from us by a sheer drop of eight hundred feet, lay brown and green fields and the inconspicuous polygonal mud hogans of Navaho Indians. In shaded recesses at the base of and part way up the cliffs were perched the remains of dwellings built by people who had lived there a thousand years before. The floor and walls of the canyon and the world they contained were graced with a sense of unreality. They seemed forever unreachable—a place, a culture, a history in which we could play no part.

The sight of the canyon and the people living there seemed to invoke a certain remorse in my companion. "For them it's all right," he said. "They have their place. But for us, there's no place left, or at least I can't find one. Do you know," he asked, "where a man can be at home?"—and left me with that question.

The next morning I was bumping along the floor of the canyon in the back of a four-wheel-drive pickup truck, watching the massive sandstone walls twist past me. Unfettered horses grazed in side canyons; cottonwood groves grew next to where the stream would be when it ran in the wet season. "What place is home?" I asked myself. The lack of a ready answer should have bothered me, but I felt only the shaking of the truck as it rattled on.

In the remoter recesses of the canyon three children were digging for water in the dry stream bed. Like the roots of the cottonwood, they knew where to look.

"What a strange view of the world they must have," I thought. "For them the universe is a narrow strip of sand and mud between massive, sky-high walls. Within the valley, each rock, each angle and shadow has its special meaning and dimension. Some are places to run to when flash floods roll down from the east, others are sites reserved for the gods. How at home they must be here —until rumors of the world above the rim, outside the canyon disturb them, and life takes them to where the earth is no longer familiar."

What answer is there to questions such as "where is home?" What answer can there be? It occurred to me then that the only solution is to be at home everywhere. But if I had said that to the man looking for a place to live, he might have thought me unsympathetic; and if I had said the same thing to the Navaho emerging from the canyon into the modern world, he might have thought me unfeeling.

Much of life these days, it is true, especially life in the cities, is incompatible with the ideal of a real home and meaningful existence. It is life divorced from the canyon, from the earth and from other men. But actually the earth is everywhere, and from it, if only we can sense it, there emanates constantly the wherewithal for man to know what he is and where he belongs. Awareness of the earth, consciousness of its proximity, of its inescapable influence—even when not obvious—presents aesthetic and psychological possibilities largely overlooked or forgotten. Each individual, in canyons or beyond, is deeply affected by his physical surroundings. If it can reach him, knowledge of the earth as reality, rock as material of the universe, landscape as momentary expression of natural process, is a rich and vital source of sanity and calm for modern man.

It is here that geology and the geologist have a contribution to make. With the geologist lies the special responsibility and opportunity of revealing the earth in all its beauty and power. The geologist is in a peculiar position. He has a foot firmly planted in each of two worlds: the modern scientific world, with all its abstraction and complex tools, and the world of the actual earth, with its primitive substance and concrete being. This dual aspect of his profession arms him uniquely, if he so chooses, to present the earth to modern man in relevant and acceptable terms. As a scientist, the geologist is part of modern society and thus speaks its language and is familiar with its problems; at the same time he is a human being in contact with the mystery of the earth and partakes of its nourishment. If geology and the geologist neglect interpretation of the earth to society, they are guilty of relinquishing what should be one of their major contributions.

Such an admonition is not idle or uncalled for. The geologic discipline at this time has reached a point in its evolution where it must make conscious decisions with respect to its identity and goals. It is a science at the brink of revolution; at the moment its character is flexible, and what it will be in the future is not yet fully determined. As the sum of its knowledge increases exponentially, due to the introduction of recent instrumental innovations, its old dogmas are collapsing and a descriptive past is giving way to an exciting era of broad hypothesis and unifying generalization. Indeed, change has been so rapid that classical and modern geologists temporarily coexist—and gaze at each other somewhat aghast.

Change is no doubt inevitable and largely desirable, but ironically, there is danger that the "new" geology will become alienated from what it seeks to know. Technical progress increasingly involves geologists in investigation of earth properties inaccessible to unaided human senses. The extent to which direct contact by the senses is being replaced by indirect contact through gadgetry, bodes spiritual and ultimately scientific disaster. What, after all, is meant by "knowing about the earth"? Are its numerical parameters and exact sequential history all it has to offer? If geologists allow too great a separation to develop between themselves and actual earth forms and materials, that is, if they cease to wander the hills and feel the rocks and dirt beneath their feet, they, like the rest of modern man, face the possibility of understanding less and less about themselves and what they purport to study—the earth.

Another consequence of advances in geology—and all science—is that they become increasingly unapproachable to the outsider. The separation between scientists and science and the rest of man goes beyond a gap in factual information. Knowledge of a way of life is missing—and this hiatus is more or less institutionalized by the structure of modern society. For instance, most formal textbooks or courses in scientific or other disciplines are, by design, impersonal. The student is enlightened factually, perhaps methodologically, but he rarely emerges with a sense of what it is like to *be* a geologist, a biologist, an economist. Perhaps this is because the author or teacher doesn't consider that his subject matter constitutes a singular way of seeing the world, or that his own particular feelings or activity are important or relevant. It is, after all, never completely an accident that a man is a geologist—or a member of any other profession or trade—if he is at least partially at home in his work. What he does is a reflection of what he is, and it permits him to view the world through a unique prism. Thus, the transmission of the flavor of a discipline is as important as its accomplishments and approaches: the struggle, the uncertainties encountered and inherent, the feelings and attitudes of the people involved. Somehow the barrier must be broken, the common human aspect revealed. This can only be done, of course, on a highly subjective basis, stemming from each man's own sense of the universe, but then all men can participate and share their human experience.

Furthermore, it is quite possible to go through a course in geology (formally or informally) without suspecting its philosophic, aesthetic or psychological potential; that is, without realizing its connection to *life*.

The aim of this book, therefore, is to suggest to the reader the relevance of the earth to his own conscious and subconscious existence and to urge him to assess its implications on a personal level. The essays that follow are in no way meant to be an orderly and rigorous description of the content of geology. Trains of geologic speculation and presentations of factual material are brought in to illustrate the nature of the subject and the state of the science. Above all, however, the attempt is to show that the earth, in and of itself, has direct meaning for every man; that human life is bound to take place beyond the rim as well as within the canyon, and that all environments have the potential of being home.

The Innocence of Rock

There is an innocence about rock. The pebble in my hand, the boulder I stand on, the cliff wall next to my face that shades me are quiet and deceiving. Air and water rush and echo against their surface, but they themselves say nothing. If in an excess of energy I hurl the pebble into the sea or pry the boulder loose and let it roll down a slope, I may feel I am the master. In determining their position, their movement, I gain a temporary illusion of power, even if secretly I know that in the long run I am less enduring than they.

That this is an illusion, however, is clear. The boulder on which I stand, the pebble that I release or hold—it is they, really, that hold me. Together with the mass from which they have been broken, the mountains and plateaus where they originated, they state the framework of my existence: permit it, define its possibilities, outline its limitations. In short, they comprise the earth, whose own beginning and growth lay in the careless accumulation of fragments such as these—cosmic pebbles and boulders swept from space some five billion years ago—and whose aggregate material musters the feeble gravity that clasps us to it and keeps us from spinning off into the void.

Balanced Rocks near Moenkopi, Arizona

Is this essential function performed in innocence, unknowing and mindless? There is a passive quality to rock that makes us think so. It suffers with seeming indifference our attempts to alter it, carve it, tunnel it, heap it up in mounds, to shape it to our will. With luck it doesn't cave in, shift or otherwise assert its presence. It is just there, ultimately important, immediately subordinate.

There are, of course, great belts around the earth—about the margin of the Pacific and through much of the Mediterranean and central Asia—where the passivity of rock is suspect, where stillness and predictability are rightly judged as momentary pauses in an almost continual round of quivering, shaking and erupting. There, rock is the enemy as well as the giver of life. A hostility emanates from it that transcends indifference of occasional moments of idleness.

But places where rock is known as fickle and places where it wears the mantle of eternity are, even within human experience, interchangeable. How short is memory; how great is our need to trust. If the flicker of time was quickened and the centuries passed as seconds, we could not ever forget the unstable nature of this globe that we inhabit. The evident drifting of its continents, the rapid rise and fall of its mountains, the swift draining of its seas like giant tide pools from one area to another, would impress deeply upon us the precarious character of our existence—and the ultimate generosity of the earth. A single lapse, a second without tenderness, and all rock would be cleansed of its living debris.

If rock is primarily our foothold on life, it is also a reminder, a link—if we let it be—of our relationship to and the reality of the rest of the universe. No less than the stars, the rocks so casually around us are the stuff of the world: their siblings circle the sun, their godparents populate all space. When the pebble in my hand returns the pressure I give it, its coolness or warmth, its quiet weight are not really an indifference. They are what is, for what else exists, really, after the pebble I hold and the boulder on which I stand?

There is love, violence, the complexity of life. But life is in-

Punta de los Lobos Marinos, Point Lobos, California

extricably intertwined with the substance at hand, materially and psychologically. The shape of civilization, beyond what it has to extract from the earth, is evolved in the shade of the cliff, the heat of the plain, its fiber and warp bent to the contour and sweep of the land around it. Even in the heart of the cities, brick is from clay, cement derived from the limy excesses of antediluvian seas. In moments of hesitation or despair such knowledge can be a measure of sanity, a route to reality. The "natural" world is the world, and the productions of man extensions of it—always—even when seemingly obscene or tortuously remote. It helps to remember.

The variety of rock—its colors, textures, topology, its contrasts with life, light and water—is a deep and endless source of regeneration: the roughness of a stone fireplace, a walk through the visceral winding of a canyon, the pebble in your mouth to keep away thirst, the rattle of shingles on a beach, chance granite that forms a curb or paving block, the sheet of marble that rises to the sky. There is no stinginess about rock! Sought out or revealed unexpectedly, apparent or disguised, it is the shape and substance of things—within whose folds we are born, on whose flanks we are passionate and follow our fate, to whose depths we eventually return. It is rock from which the seas were squeezed, the atmosphere expelled, the accident of whose mass keeps our molecules close rather than flying, wasted, into space.

Rock is innocent. It is innocent in that it presses its case so little. It is innocent in that we—its children—have not burdened it with the guilt we attach so easily to our own lives, innocent in that we free it from the necessity of justification.

But if rock is innocent—no more or less so than an animal or a plant—as its offspring, doesn't this quality devolve, ultimately, upon us too?

The Geologic Experience

If rock is innocent, with all that that innocence implies, a significant question to ask is whether geology, as the scientific discipline that seeks knowledge of rock and of the earth, is particularly equipped to transmit the nature of the geologic experience to the rest of mankind. That is, whether geology—more pertinently, geologists—are capable of perceiving and communicating the nature of the earth-human relationship. A consideration of what is the geologic experience, the history of geology, and some perhaps forgivable generalizations regarding the personality of geologists provide an answer.

What constitutes the geologic experience depends largely upon the attitude of the individual who beholds the earth—the ground rules he establishes. Their substance is quite arbitrary: they may be formulated for definite purposes, or stem from ingrained, unquestioned beliefs. Until the eighteenth century, the forms and processes of the earth were considered by most to be manifestations of the will of unknowable deity, and as such were free to be capriciously benign, malicious, orderly or chaotic, generally beyond the scope of human understanding. In short, the ground rules were such that "anything is possible." Detailed, systematic investigation was therefore meaningless, except, perhaps, to determine the prevailing disposition of the gods: what might please them, what they deigned to offer man and what threat they posed. In the light of this attitude, every crag, knoll or spring was fraught with significance, colored by portent beyond the visible. The earth was something to love, to fear, to keep careful watch upon. The sin most likely to incur catastrophe was indifference.

Modern earth knowledge and the beginnings of geology as a science originated with the establishment of a different set of ground rules, those formulated by James Hutton in the eighteenth century, subsequently known as the Doctrine of Uniformitarianism. According to this doctrine, nature and the earth are not arbitrary or capricious, but follow knowable, predictable and unchanging patterns of cause and effect. Earth processes, it was claimed, are understandable—if we but have the patience and ingenuity to understand them—and from earth forms evolved by earth processes, earth history may be inferred. Thus, the mountain, the valley, the curiously balanced rock are not gods or their imprint but the result and record of earth movement and erosion, a statement of the present, momentary balance achieved between creation and decay.

If this is so, then the deep canyon may be carved imperceptibly by the flowing river, and the continents and mountains be thrown up gradually by infrequent earthquake and intermittent volcanic eruption. And if *that* is so, then simple arithmetic comparison of rate of process with the dimension of accomplishment reveals that the world of today must be the result of thousands of billions of yesterdays.

What can such a number mean? How quickly the assumption that the earth is susceptible to rational interpretation leads to conclusions as incomprehensible, as inconceivable as the motives behind the dictates of deity! Who, after all, can really grasp the length of geologic time? Analogy offers momentary illusion of comprehension. If the time that has passed since the creation of the earth is represented as a cliff one mile high, then all of historic

Great Sand Dunes National Monument, Colorado

time may be found in the uppermost tenth of an inch, and a single man's existence encompassed within less than the thickness of the finest hair.

Indeed, within the span of such time we can again conclude that "anything is possible." It is not the same "anything is possible" that implies imminent, unpredictable and unexplainable events, but rather an "anything is possible" that attributes to nature the vastness and canny ability to create not only everything we see, but —from within the bottomless pool of her own resources, employing endless combinations of perhaps reasonable, ultimately knowable mechanisms—anything we may dream or dare not dream of.

Thus, the immediate uncertainty of religion was exchanged for the long-term uncertainty of science. They are not equivalent, however, for in opting for science, man's ego lay claim to the possibility of rational understanding and, therefore, to the potential of controlling his surroundings. Not that control was not sought before the advent of science; to that end the arts of supplication, persuasion, propitiation—attempts at anticipating and fulfilling the wishes of the gods—were highly developed. There was a conscious striving to sense and participate in the earthly and divine harmonies. Control through the rational approach was more direct —involving manipulation and direct human intervention in natural processes—and in many ways was more obviously effective.

At first, however, the results of scientific investigation of the earth seemed to pose a cultural affront, directly contradictory to the Judeo-Christian concept of a world centered about man and his God. If the eons of geologic time were accepted as reality, then the formation of man through modification of more primitive life—through a process of gradual evolution—became a serious possibility, as unpleasant a thought as that of the earth as a modest-sized planet of a modest-sized star located part way towards the edge of an entirely undistinguished galaxy.

The relegation of man to a minor and insignificant position in the universe was a harsh challenge to religious pride and dogma, a challenge increasingly successful in diminishing ecclesiastic strength and authority. If, however, theological doctrines of man's centrality and uniqueness began to crumble, his own internal, non-rational opinion of his importance did not. Science might reveal man as minor and insignificant, but that was something it was best not to think about too closely. More important was the fact that science led the way towards man's dominance of the proximate universe, which he then proceeded to bend to his will as fast as he was able.

Thus, even if the detail and paraphernalia of Judeo-Christian religious ceremony fell away, its viewpoint as to man's relationship to things about him remained. Still convinced inwardly of his fundamental centrality, and armed with new and increasingly powerful technologies, man began to hack away at his environment with unrestrained vigor. Apparently passive, the earth yielded streams of oil, coal, metal, fertilizers, salts, building materials, chemicals, water, and what have you, in seemingly endless abundance. If geologic, astronomic and biologic research revealed the universe as large and seemingly infinite in time, then small and short-lived man felt partially compensated for this by the help they gave in providing him with the tools to alter his surroundings and to make himself comfortable. This aspect of science was acceptable and desirable; experience of the earth, the geologic experience, was now one of discovery and exploitation—surely preferable to what he somewhat inaccurately remembered as previous subservience and fear.

But in its own way, the earth has begun to protest this attitude. Its material resources, though generous, are far from inexhaustible—another fact which simple arithmetic reveals. Its maxims are prosaic but clear: waste not, want not; you will reap what you sow. The earth will remain, but as part of its delicate coating of life we may not. We may modify our environment only so far without initiating chains of irremediable disaster; water, soil and atmosphere may be tampered with only so much before they become unusable.

More and more the earth itself begins to dictate what must be the geologic experience, the human experience. It must be an experience of appreciation and reverence, of use and replenishment. The modern analysis of science and the awe of primitive man must combine to show the way. Man must learn to view himself as part of rather than apart from the earth. As he does, he will begin to experience grandeur and satisfaction totally beyond what may be achieved through "exploitation." Earth resources are spiritual and intellectual as well as material. Earth processes employ and reveal the mechanisms and economy of the universe. Earth forms and substance offer themselves to us in an infinite number of ways.

The rules of the geologic experience, the rules of a true geology, must stem from ethics as well as reason, beauty as well as production. The Doctrine of Uniformitarianism proposes that intelligent observation of current natural processes permits an orderly and reasonable interpretation of the past. Each act is father to its consequences. With respect to our own relationship to the earth, this same doctrine may well suggest that unless our existence achieves harmony with nature, there may, for us, be no future, and the geologic experience will have to unfold without its human participants.

Geology—a Chance to Understand

The geologic experience must be a human experience; that much is vital to man's continued existence on earth. It therefore behooves the geologist to ask what is the relationship between geology and the geologic experience, whether the "real" geology, geology as it has evolved since its inception and as it exists today, holds the promise of the type of earth knowledge that will contribute to man's eventual reconciliation with nature.

Because of geology's stormy past and uncertain future, with the resultant necessity for much professional soul-searching, there is the chance that the prognosis may be favorable.

Geology is perhaps the youngest of sciences; the name itself has received general acceptance only within the last two hundred years. While it has made major contributions to the fund of general knowledge and philosophic outlook, it has, at the same time, become highly dependent on the data and methodologies of its sister disciplines. Intertwining of sciences is not surprising. What is surprising is the notoriety, derision and obscurity that have plagued geology, the reasons for which not only explain its weakness but imply its strength.

As already suggested, the major contributions of geology to

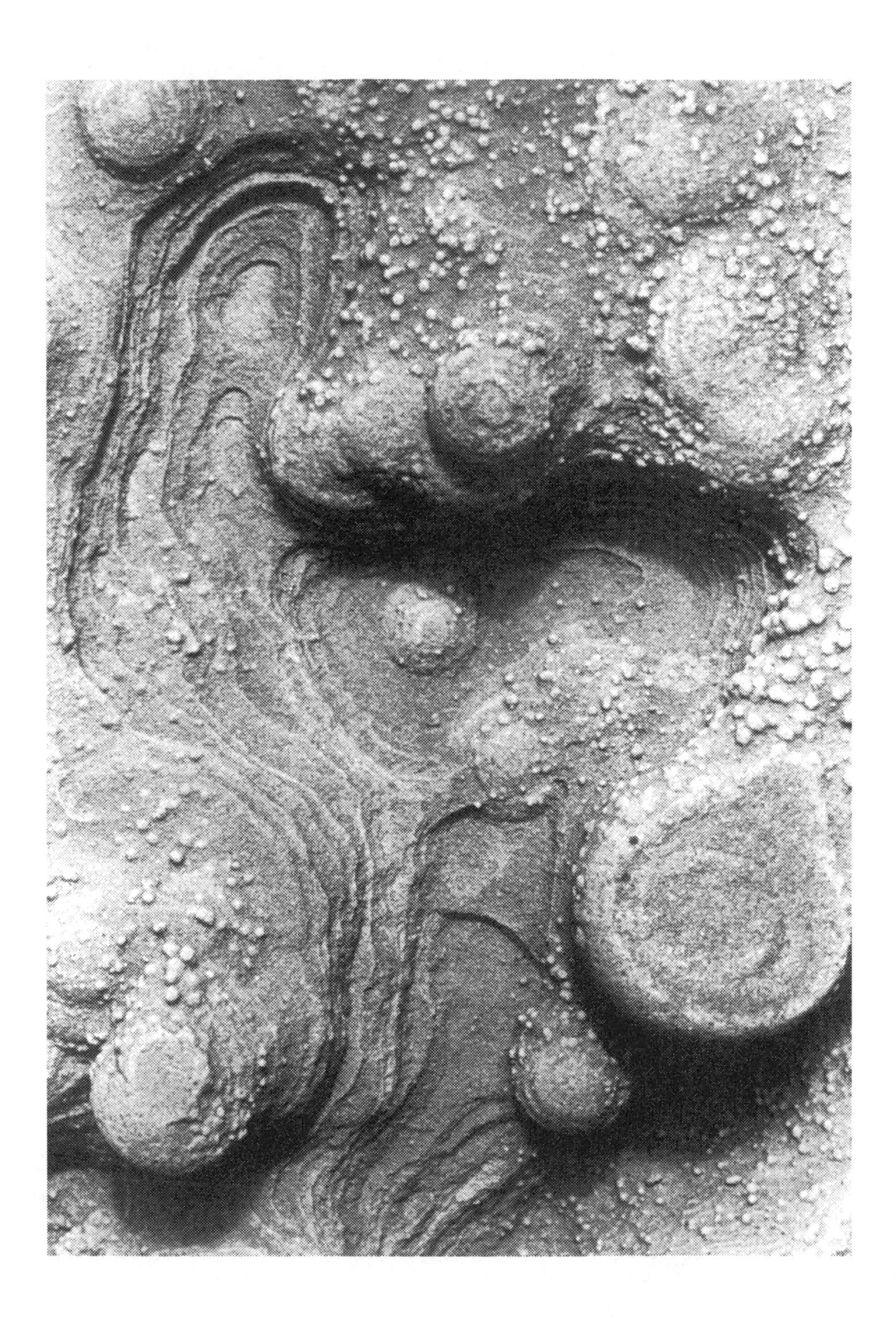

general knowledge include the concept of geologic time, the Doctrine of Uniformitarianism and substantiation of the theory of evolution. Those of conservative bent, to whom these ideas were anathema, found geology an obvious target for their response—and the response was bellicose and resounding: geology was the science of the Devil.

Early geologic contributions of a *practical* (as opposed to philosophic) nature—attempts to understand the genesis and location of ore, the distribution of ground water, the effects of erosion on soil, et cetera—met mixed acclaim and scorn. Suspicion of the scientist as an interfering, ineffectual intellectual was coupled with glad welcome at successful prognostication. In the United States, settlers of new territories became well acquainted—whether favorably or not—with the existence of geology and geologists. The figure of the latter was easily recognizable (at least his image was clear in the public mind): a mountain-climbing canyon-crossing loner armed with a portable assay kit and a good hammer—and probably the pin of some eastern college on the underside of his lapel.

Today all that has changed. The general public doesn't know what a geologist is or does (he probably has something to do with rocks or oil—or is it pottery?), and the geologist himself wonders "what is geology?"

As in most sciences, the early days of geology involved accumulation of data and occasional attempts to summarize and interpret this information. Terminology proliferated; the cataloger ruled. Economically valuable deposits of ore, oil and non-metals were rapidly located and opened for exploitation. Then, with time, geologic concepts and methods were shaped and sharpened by the increasing sophistication of chemistry, physics and mathematics. Technology replaced terminology. Attempts at mathematical analysis and experimental duplication of natural processes multiplied. An era of brilliant laboratory work in high temperature, high-pressure synthesis of rocks and minerals was initiated at the Carnegie Institute in Washington, D.C., that permitted critical ex-

Rock detail, Zion National Park, Utah

amination of volcanic and allied processes. Engineering and metallurgical advances threw light on how materials react to stress, and theories of rock deformation and mountain building resulted. The discovery of X-rays made possible study of the internal structure of materials. Earlier, the invention of the polarizing prism by Nicol and its use in the petrographic microscope allowed examination of the detailed fabrics within rocks. Description of landforms yielded to attempts at quantification of process. The studies of fossils, ancient life environments, the deposition of sediments and their stratification and correlation, all donned the capes of statistics and multivariate analysis. Development of radiometric techniques put numbers to geologic ages; isotopic tags revealed the distribution and migration of elements within the earth's crust. And so on and so forth. The lone geologist on the mountainside suddenly found himself accompanied by a thousand pack mules and ten thousand colleagues demanding information and cooperation.

Times had changed. The era of the "general" geologist who had wide, speculative interests collapsed into an age of specialization. The "general" geologist was to emerge again, but as a different creature. He became a synthesizer of other people's research and was viewed suspiciously by his colleagues. One couldn't, anymore, be a geologist; one was a geochronologist, a paleobotanist, a metamorphic petrologist. Anyone who could claim critical understanding of all these fields was either a genius or a fake—and in either case a dubious character.

As specialization increased, the nature of geologic training changed. Indeed, it became better *not* to be trained as a geologist, but rather to be a biologist, chemist, physicist or mathematician with an interest in geologic problems; that is, to coin a phrase, to become a "geospecialist." The relationship between the geologist and the "geospecialist" was, from the geologist's point of view, like that between the servants and masters on Gulliver's island in the sky. On that floating island, members of the ruling class were so out of contact with reality that, as they speculated and theorized, they had to be gently led about by their servants—

so they wouldn't stumble—and patted on the mouth with bladders—to remind them to eat or speak.

Mistakes committed by non-geologically trained "geospecialists" (the application of their talents to meaningless samples and problems, for instance) were inevitable and reassuring to the ordinary geologist. Gradually, however, the symbiotic advantages of cooperative relationship became appreciated and accepted by both.

In the decades since World War II, the physical area of study encompassed by geology has expanded tremendously: from the continents to the ocean basins, from the crust to the depths of the mantle, from the earth now to the moon and soon, no doubt, to the planets. Studies of the earth's magnetic field, the origin of the atmosphere and sea water, the origin of the earth itself, blur the margins between traditionally geologic domains and those of meteorology and astronomy. Geophysics, oceanography, vulcanology combine with the sciences of the air and outer space in the ultimate grand study of the origin of everything—cosmogony. The humble gentleman picking with his hammer at the remains of an ancient fern, a cluster of crystals, the lithified shell of a clam, has been transmuted into a likely candidate for the first civilian to undertake travel through space!

In an attempt to emphasize the rise of the "geospecialist" and the decline of the classical geologist, there has been a move of late to relinquish the term *geology* in favor of *earth science*. This, I think, would be a mistake. Although the apples on the lower branches of the geologic tree have been picked, the higher ones are still there. Geology is still distinct from other sciences of the earth. It is true that one of the results of the exponential acceleration of knowledge and technique is that many long-accepted geologic generalizations no longer stand up. Indeed, some of the fundamental assumptions upon which long trains of geologic thought have been based are turning out to be invalid. It used to be almost axiomatic, for example, that continents and ocean basins were permanent, stable features of the earth's crust; the mounting evidence for continental drift (which will be discussed farther on) suggests they almost certainly aren't. It used to be held that the earth is cooling from an initially molten state—with widespread implications for the origins of volcanic activity and the building of mountains. But the earth may have formed as a cold body, which has since been heating up. These revolutions in geologic thinking, however, do not invalidate the science. Rather, they force upon it a re-examination of its theory, a re-examination of its basic data. This work is most certainly that of the geologist. Who else can undertake the mapping, remapping, and re-remapping of complex terrains? Who else is as familiar with all the possible implications of the new information?

The lone geologist on the side of the mountain is resurrected. The geologist's love of the outdoors and his intuitive skills are not yet replaceable by the computer. Indeed, the emphasis on laboratory and experimental technique has, in one sense, truly hampered geology, for those questions too complex to be treated by such methods have been neglected and derided. Advanced as the allied sciences have become, many geologic phenomena (like many life phenomena) require science that has yet to be invented. For instance, it is easier experimentally to create crystals by cooling a saturated liquid than it is to grow them in the heart of solid rock. That the latter happens geologically is certain; because it can't be duplicated experimentally the process is relegated to a position of minor importance. The failing, however, is that of the experimenter, not of nature.

Furthermore, there is the necessity of translating the geologic experience to the rest of humanity. Because traditional geology was typically non-mathematical, it could be appreciated more readily than other sciences by the lay public. Today the same public has come to accept the existence of an impassable barrier of incomprehensibility between science and themselves; indeed, they consider the barrier the hallmark of science and encourage its growth. Science is welcomed by too many as a new god to which responsibility may be abdicated. In this setting, traditional, non-mathematical geology appears as a backward child, even a simpleton. The common secret thought of the layman, perhaps, is "probably even I could understand it. So how significant or scientific could it be?" But traditional geology and its modern counterparts, the areas within geology that are non-mathematical and intuitive, have a great gift to offer: an understanding of our environment together with the chance, so rare today, of an intelligent and intimate involvement with the natural world. Together with biology and astronomy, geology leads into the universe that is beyond, yet within and surrounding man. It offers a unique opportunity for perspective and a grasp of reality, it retains and renews its vitality and meaning as befits the youngest of the sciences.

Through the Eyes of Man

In science and in much philosophy there is an insistent search for natural patterns that will support man's hopes and allay his fears. Out of the metamorphosis of the caterpillar there arises not only the butterfly, but also thoughts of reincarnation and the possibilities of life after death. Each phenomenon of nature that can be interpreted in an encouraging manner is welcomed with acclaim; those that seem contradictory or perverse are popularly ignored or belittled. Even more sought after than isolated facts are general laws or tendencies within nature from which man's destiny may be read favorably. When such laws are outlined they are widely hailed and considered as significant advances of man's knowledge—which indeed they are. These "encouraging" laws, however, have a weakness less likely to be present in "neutral" or "discouraging" laws: being so sought after they may be "found" where they don't really exist.

Consider the search for cyclicity in natural processes. Cyclic concepts are tidy, with all corners tucked in. To be part of a cycle is to be part of eternity, to hold a guarantee of immortality. Cycles embody the excitement of motion together with the security of going nowhere: the wheel turns, but that is all. Man, for instance, considers himself born of the dust to which he will return, a thought from which he evidently receives some comfort—perhaps because of its hint of renewal out of decay, suggesting that the end of life is close to its source.

It is not surprising, therefore, that closed cyclic chains of events have come somehow to seem aesthetically complete and fit our notions of how nature should behave. Equally without surprise is the fact that cyclically framed explanations of natural phenomena, being ardently sought for, sometimes escape close scrutiny.

Almost two hundred years ago, James Hutton, to whose inspiration geologists must turn again and again, envisioned the workings of the earth in terms of such a cycle. As he saw it, through the interplay of earth processes and earth products, rocks are born, live, die and are then reborn. Consider for instance the lava that pours from a volcanic crater. The lava, molten rock material from within the earth, after working its way to the surface, flows down the sides of the volcano and soon hardens to rock. Such rock, as is any rock that forms from molten material, is considered *igneous*. Exposed as it is to the elements, the hardened lava will eventually be decomposed and broken down into loose sediment. The sediment, moved about somewhat by ice, wind and water, will come to rest and perhaps be cemented to *sedimentary* rock. With the passage of enough time, there is good chance this sediment or sedimentary rock will be caught in a downfold of the earth's crust, be depressed gradually into regions of high temperatures and pressures, and be altered to what is known as *metamorphic* rock. The metamorphic rock, in turn, may in places encounter regions of such intense heat that it will actually melt. If this happens, these newly created pockets of molten rock may work their way back to the surface and be poured out as lava in another volcanic eruption, which, when it cools, gives rise to a new generation of *igneous* rock, and thus completes the cycle.

(Overleaf) Arches National Monument, Utah

The "rock cycle" presented in this manner was indeed an ingenious summary and synthesis of natural geologic processes. Vulcanism, rock decay, erosion and, by inference, the sinking of oceans and the growth of mountains no longer had to be viewed as isolated mysteries. Instead, they fit into a dynamic and comprehensible grand design. Such was the grasp the extraordinary Scotsman had of the world about him.

The world about him, however, was also the world of the late eighteenth and early nineteenth centuries. That the rock cycle was a brilliant unification of seemingly unconnected phenomena was not enough; it had didactic possibilities that could not be neglected. The rock cycle was soon seen as revealing "the wisdom that presides over nature . . . proving, that equal foresight is exerted in providing for the whole and for the parts, and that no less care is taken to maintain the constitution of the earth, than to preserve the tribes of animals and vegetables which dwell on its surface." That is, it "is a system of wise and provident economy, where the same instruments are continually employed . . ." So said John Playfair in 1802 in his *Illustration of the Huttonian Theory of the Earth.* In other words, the rock cycle had purpose and, moreover, was efficient. That is, in short, it was moral. Most important, it testified to order within nature, reason in the universe, and could be cited as a comfort to insecure man.

Whether or not moral, the rock cycle still provides the usual foundation for an integrated approach to geology. In diagrammatic and written outline it is found in the introductory chapters of most if not all texts, usually in its original circular form (with the recent exception of *Physical Geology*, by Longwell Flint, and Sanders). But is it really a cycle? Do all these processes and materials really go nowhere, except back to where they started? The answer is no—the processes do go somewhere; the cycle is not really a cycle, it is more of a spiral or a complex helix.

The rock cycle was conceived at a time when there was no

knowledge of the major structural divisions of the earth (the earth as we now know it consists of a relatively thin crust, a thick mantle, and an even thicker core), and all the integral processes of the rock cycle were treated as a more or less closed system originating in and restricted to what is currently called the earth's crust. But they aren't. In the operation of the real earth, the source of most volcanic material lies far beneath the more superficial crustal levels where erosion, weathering, sedimentation, burial and metamorphism take place. These latter processes occur within the outermost thirty miles or so of the earth's four-thousand-mile radius. But most rock melts originate not in the crust but in the underlying mantle, and from there make their way upwards towards the surface. Once the molten liquids arrive in or on the crust, the igneous rocks they form may take part in the crustal processes of the rock cycle. But the reverse procedure—the return of crustal materials to the deeper mantle—is a difficult and at best temporary proposition.

A closer look at the nature of igneous materials helps to understand the inadequacy of the circular concept. Most volcanic products and other materials of undoubtedly igneous origin are composed of the minerals plagioclase feldspar, pyroxene and olivine, and the rock they form is classified as "basalt." No one has ever seen rock from the earth's mantle, but rocks of basaltic composition, according to most educated guesses, are what would result from local mantle melting. It does seem reasonable, therefore, to say that the source of most basaltic, and therefore most igneous material occurring in and on the crust, is the mantle. (Furthermore, studies of earthquakes associated with oceanic vulcanism confirm that the lava has worked its way up from great depths.) Igneous material, however, that originates within the rock cycle through melting of deeply buried sediments, is *granitic* in composition—quite distinct from mantle-derived basalt. That is, igneous material that results from recycling of older rock is clearly different from molten material *introduced into the rock cycle for the first time.*

Thus the rock cycle cannot be a closed system going nowhere for, in the course of time, there is a continual transfer of basaltic material from the earth's mantle to its crust. In fact, it is through this upward and outward migration of molten mantle material that the crust formed in the first place and has since attained its present thickness.

In its early stages, at some short time after its initial accumulation, the earth may be pictured as having been a more or less solid, homogenous mass. Its subsequent division into a heavy, metallic core, an immense intermediate iron-magnesium rich silicate mantle, and a thin outer potassium-sodium rich silicate crust has been due to an eons-long, gradual gravity separation which is still going on: heavy, compact atoms have worked their way downwards towards the earth's center; lighter larger atoms have diffused slowly upwards towards the surface—like a very slow unmixing of cosmic oil and water.

Indeed, if the rock "cycle" is circular rather than spiral, the earth would not be at all as we know it. But as a subcycle within a larger spiral that embodies progress and evolution rather than endless repetition, Hutton's idea remains masterful and invaluable.

What about other cycles seen in nature? Are they too misleading or are they helpful? The hydrologic or water cycle, for instance, seems valid enough at first glance. Water may follow a lengthy and devious route, but after having been precipitated from the atmosphere, it will eventually find its way back up again. It may return almost immediately through evaporation, be absorbed by plants and then breathed out again, or run through streams or the ground-water system to lakes and oceans before it is once more vaporized. However, like the rock cycle, the hydrologic cycle is not really completely closed: for water (or steam) is continually expelled from the interior of the earth in the course of volcanic eruptions and is added to the cycle, gradually, over the length of geologic time, swelling its total volume. Indeed, the ultimate source of all water at or near the surface of the earth is to be found in this "dewatering" or "degassing" of the earth's interior. Thus the hydrologic cycle, like the rock cycle, is more accurately a spiral.

The geologic cycles mentioned this far have one attribute in common: in spite of their errors they have been a helpful way of looking at things; there is no doubt that much more has been gained than lost through their formulation. Such unambiguous benefit, however, cannot be invoked for all geologic cycles. Consider what is commonly called the "erosion cycle"; an attempt to describe a circular sequence of stages called "youth," "maturity," and "old age," that an area undergoing erosion supposedly passes through.

In youth, more-or-less flat elevated land, such as an uplifted lowland plain or former sea floor, constitutes what is defined as "original surface." This raised surface, exposed to the ravages of weathering and the elements, is gradually replaced by surface that is the result of erosion. First gulleys, then well-developed streams carve the area into a system of valleys, restricting the original surface to divides or highlands between the valleys. Then, with the passage of time, the flat divides shrink to a network of crested ridges between the eroding streams, and the last of the original surface is gone. When this happens the area under consideration is said to have passed from "youth" to "maturity": youth is characterized by the presence of original surface; maturity by its absence.

In their early phases mature landscapes still retain most of the elevation gained in the original uplift, even though the original surface is lost. The effect of further erosion, then, is to gradually lower the elevation of the area by removing its bulk and substance. The end result, therefore, is a more or less flat, featureless plain reduced to an elevation approaching sea level. This end-stage erosional surface is called a *peneplain,* which means "almost a plain." When peneplanation has been achieved, the area is said to have entered "old age."

Old age for landscapes, however, does not signal death but

rather the possibility of new youth. Peneplains, as low flat surfaces, require only rapid uplift to become high flat surfaces; that is, to become new original surfaces upon which erosion can begin to operate again.

The erosion cycle as described, therefore, consists of a sequence of erosional stages that, when they have been accomplished, are restarted and re-energized by regional uplift. There are, however, a number of difficulties with this concept. First of all, for the erosional part of the cycle to progress smoothly—from youth to maturity to old age—the area must remain stable, for any tilting, warping or uplifting during erosion would interrupt the cycle and create new original surface before the old original surface and elevation were completely removed, thus superimposing two or more erosional stages, one upon the other. In actual fact, the erosional pattern of most regions *is* complex, indicating interruption is the rule rather than the exception. Secondly, the uplift that converts an area of old age into one of youth must be relatively rapid so that erosion cannot accomplish significant results during the uplift—otherwise uplift would not result in unsullied original surface. In some cases uplift may indeed be relatively rapid, but there is abundant evidence to suggest that in many other cases uplift is both gradual and intermittent. Lastly, as a *coup de grâce,* peneplains, the ideal end-stage of the cycle and forerunners of youth, are not actually found anywhere—except as dissected remnants that can only suggest their former existence.

In other words, the nice sequential synchronization between crustal deformation and erosional stages described within the erosion "cycle" seems more fancy than fact. Restricted local areas with the appearance of youth, maturity and old age are common enough, but there is no particular assurance that youth is prelude to maturity, or maturity the predecessor of old age.

Furthermore, the assumption of the erosion cycle as commonly described is that the net result of erosion is the wearing *down* of high areas. There is much evidence, however, that in arid

Junction between the Rio Grande and the Rio Frijoles, New Mexico

as opposed to humid regions, the processes of erosion result in a wearing *back* rather than a wearing *down*. If this is so, then youth, maturity and old age would have to be completely redefined.

All in all, therefore, it seems better to consider youth, maturity and old age as useful descriptive terms: youth (in humid areas) referring to extensive elevated undissected flatlands; mature regions as those displaying well-developed valleys and ridges with little or no high flat areas; and areas of old age as low undulating surfaces where erosion is relatively inactive. Used in this manner, these terms are helpful summaries of regional topography; used with cyclic or sequential implications, they are seriously misleading.

Thus, the attempt to apply the aesthetically satisfying concept of *cycles* to natural phenomena has, in one instance, been a failure, and in two other instances been inaccurate. In the case of erosion, the cycle collapses into what may be termed physical or descriptive knowledge—knowledge of the way something is at a given moment without suggestion as to its future state. For the rock and hydrologic "cycles," the knowledge is "historical"; that is, the processes "go somewhere."

In what direction is this "somewhere"? The answer is towards an earth that is increasingly structured and less homogenous. The progressive division of once uniformly distributed matter into more and more distinct layers or shells that are *different* from one another may be appropriately called *differentiation*. Thus, the development of the core, mantle, crust, hydrosphere (and atmosphere) are a result of the differentiation of the early, primitive earth. Furthermore, it has been suggested that still earlier, at the very beginning of the universe, all matter was "collapsed" into a superdense uniform aggregate in which there was no distinction between different chemical elements or even separate atoms. Passage of time and the operation of cosmic processes, then, differentiated this mass into atoms, elements—and the possibility of stars and planets—and us.

Perhaps, therefore, differentiation or progress is a more valid generalization about nature than is cyclicity or stasis. Consider man himself. His thoughts do not really stop at contemplation of his return to dust: for his children he hopes for an increasingly better world; for himself he gropes for belief in life after death. Perhaps, then, man's higher aspirations and hopes, rather than his search for security, are a more accurate reflection of the reality of nature. Or perhaps not. Perhaps any attempt to force nature into a mold likable or even comprehensible to man is bound to failure. As men we must see nature through man's eyes. That, I am sure, is both our advantage and our limitation. But beyond the shortcomings of our senses and our psyche, there seems little point in adding unnecessary philosophic circumscriptions. Physical knowledge, for example, knowledge that goes nowhere, is different from, but not necessarily inferior to, historical knowledge, knowledge that goes somewhere. In some cases, physical knowledge may be prelude to historical knowledge; in other instances, it may be all there is to know. Man's anxious and eager application of "satisfying" or "reassuring" concepts to natural processes is, in a way, only a manifestation of his own fears and impatience. Progress or security, circles or spirals, may often as not be irrelevant. We should leave ourselves open. Perhaps nature is as we have never dreamed.

The Language of Geology

First encounters with geology are often those of despair. Traditionally, the emphasis of material designed for the beginner has been the enumeration, in great detail, of all the diverse features that mark the face of the earth. Having hoped to learn about the world around him, the neophyte finds himself, instead, caught in a morass of terminology. When he escapes he is not likely to wander back. If he did, he might discover that underneath the bog there is a good substrate of theory and process to stand on. He might even come to appreciate the exotic flavor of the geologic language.

Actually, there has been solid, historic reason for this emphasis on words. The surface of the earth is varied and complex; its components are not simple—they have may subtleties, variations and implications. In the early days of a science, when interrelationships are not well understood, unnecessary distinctions are often made and names proliferate, seemingly without end. Geology at the moment has one foot in the past and one in the future. It is a rare geologist who himself is not guilty of inventing one or two more names for things that seem new to him; at the same time, the driving forces, the fundamental processes that unite disparate phenomena are appearing.

The innocent who is enduring his first encounter with "the science of the earth" may have little patience for these explanations. There is a genuine drift towards sparing the casual student all the terminology that geology has to offer, but unless he has a particularly orderly streak in his character, he may still feel inundated.

It is important, however, to understand that the character of geology is not arbitrary; geologic vocabulary has a definite function, even today. That it acts as a barrier between geologists and the public is unfortunate, but its justification is real.

What alternative could there be? To give rocks, minerals, landforms, fossils or stratigraphic formations formularized names consisting of numbers or letters would be quite impractical: they would be hard to remember and would produce few or no associations in the mind of the hearer. When a geologist says basalt, brachiopod, High Falls Shale or subsequent stream, he attempts to convey a complex parcel of information. For instance, the formation name High Falls Shale signifies a mass of rocks of certain colors, mineralogies, chemical compositions and textural characteristics that (1) may or may not contain any or particular fossils, (2) has a certain range of thicknesses and a specific areal distribution, (3) suggests a period in geologic history during which it was formed and subsequent periods when it was deformed, and (4) was originally described near the town of High Falls, New York. It could, of course, be referred to by age, location, etc., as "SilUSNYhf3," but that would make conversation rather difficult.

Other sciences, chemistry for instance, have been able to abbreviate the names of materials and state relationships in terms of formulas and equations. But geologic materials are too complex and geologic processes too devious or obscure. Abbreviations, formulas and equations would hinder rather than help.

The use of names produces its own complications, however, aside from difficulties of remembering. Names can obscure as well as clarify. Labels often signify different things to different people and discussion may never get past attempts at definition. If definition is ignored, discussion continues based on misunderstanding. A classic geologic example is the term "granite."

Granite was early defined as a coarse-grained, crystalline rock composed principally of orthoclase and quartz, commonly used as a building material. The meaning was descriptive and unequivocal. Subsequently, theories of how granite formed began to evolve, and these opinions crept into the definition.

In the eyes of early workers, granite was considered to be a precipitate from some primeval ocean and thus sedimentary. It did, after all, have a crystalline texture (a texture of interlocking grains) similar to that of salts formed from evaporating sea water, and often underlay clearly sedimentary material such as sandstone, limestone and shale. Further field observation revealed that at some localities granite did not merely underlie sediments, but seemed to have risen and intruded through them and to have baked or altered the material with which it came in contact. The conclusion drawn was that granite could not have been a sediment, but must have been a liquid, high enough in temperature and subject to sufficient pressure to melt or force its way into other rocks and in the process to inflict severe changes upon them. Granite, the rock, was evidently the result of the cooling and crystallization of this liquid. Quite logically, the definition of granite was adjusted to incorporate this opinion as to its origin. Granite was now a "coarse-grained, crystalline *igneous* rock composed principally of orthoclase and quartz . . ." The word *igneous* signified its liquid or "magmatic" parentage. A mass of experimental field and laboratory data soon supported this opinion as expressed in the definition—and it seemed indisputable.

But disputed it was. In the first half of this century a school of thought arose that attributed granite to *metamorphic* processes. That is, they held that granite was the result of profound alteration of pre-existing rocks—at high temperatures and pressures—without, however, actually melting them and producing a liquid. And a body of field and laboratory evidence was assembled to support this opinion.

It was at this point that the confusion over the meaning of the word "granite" developed, and that the shift in its definition from descriptive to genetic (containing opinions as to its mode of formation) became an obstacle to meaningful discussion between the two schools of thought. Obviously, it was claimed by those who favored an igneous origin, those who talked about a metamorphic origin for granite hadn't read its definition; they must be talking about something other than granite. Perhaps they meant rocks that were granit*ic* rather than granite.

The confusion spread further. The major occurrences of granite (whatever the latter is or means) are in large bodies called "batholiths." Batholiths are granitic bodies that outcrop over more than forty square miles (an arbitrary but accepted figure), have no observable bottom and are typically found in the interior of what are or were great mountain chains. The preceding definition is quite descriptive. However, in much of the literature, and to most geologists, the term batholith connotes igneous, based on the assumption that there is no dispute as to the origin of granite: since granite is igneous and batholiths are made of granite, then batholiths are igneous too. This confusion over the meaning of batholith is particularly unfortunate, for solving the problem of the origin of granite must rest largely on the nature of its field occurrence, and many investigators are immediately biased when it is said to occur in a batholith.

A compromise solution that permits bias and description to coexist comfortably is the gradually increasing use of modifiers before nouns; terms such as *igneous* granite or *metamorphic* granite are the result. The word "batholith" is often preceded by adjectives that are descriptive and neutral: intrusive, concordant, discordant. Unhappily, even these adjectives have developed connotations that vary according to the philosophy of the user.

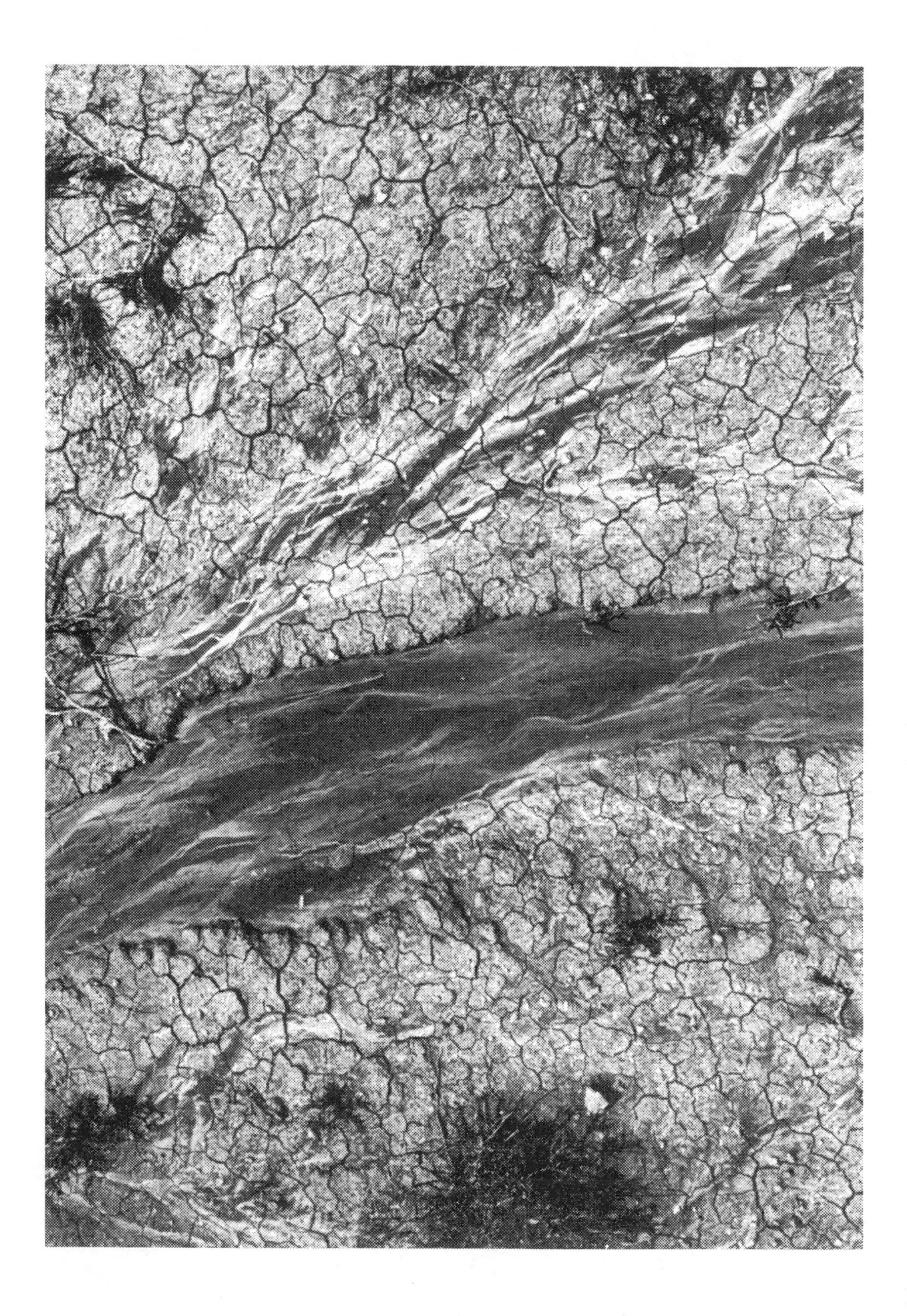

These kinds of semantic problems, frustrating but human, are an integral part of geology. In a sense they are valuable, for they are the type of failing that dispels awe and makes the science approachable. It has, so to speak, feet of clay.

Geology's abundant terminology stems also in part from the difficulties in seeing that a given phenomenon may be a special case of a more general class of things. For example, most igneous rock names now connote certain ranges of texture and mineral composition, and almost all occurrences can be fitted, without too much distortion, into one of only several dozen categories. In earlier days, however, each specimen or body seemed dramatically different from all others, and the habit arose of naming rock types after the localities where they occurred. Thus, names like ouachitite—after rocks from the Ouachita Mountains—*boston*ite and *cordtland*ite proliferated. Some of these names have stood the test of time as representing really exceptional rock types, but many collapsed into more general names modified by appropriate adjectives: thus, *quartz* monzonite for adamellite, *quartz* latite for dellenite, *hypersthene* granite for charnockite.

This reduction in the number of rock names through the use of qualifying adjectives may seem desirable, orderly and logical. But there is a built-in danger in this procedure. Until rock genesis is well understood, significant differences are hard to separate from insignificant ones. In other words, an orderly and reasonable nomenclature and classification must not be constructed at the cost of obscuring important peculiarities of some categories.

To take a case in point, substituting the term "hypersthene granite" for "charnockite" suggests that the latter is just granite containing the mineral hypersthene, and immediately plunges the rock into the whole controversy over the origin of granite. Rather than do this, it is clearly better to retain the discomfort and unfamiliarity of the name "charnockite." Charnockite presents enough problems without unnecessarily shouldering the burdens of others.

The question of terminology may also be carried from the

Detail, Ghost Ranch, Abiquiu, New Mexico

realm of practicality and logic to that of aesthetics—the ultimate practicality. In arid regions of the American Southwest and West, erosional remnants of layered rocks—more specifically sedimentary rocks, lavas and ash beds—are named according to their size and inclination. Large, flat tablelands underlain by horizontal layers are called mesas; mesas that have been reduced so that their flat tops are gone or almost gone are called buttes. Inclined layered rocks that erosion has accentuated into ridges are called cuestas if the tilt is gentle, hogbacks if the tilt is nearer to vertical. Surely the richness of language would be reduced if mesa, butte, cuesta and hogback were replaced by some duller, more "orderly" system of nomenclature.

It may just reveal the peculiarity of my own mind that I enjoy the vocabulary of geology—or the fact that I have mastered some of it. But I should be sorry to see it exchanged for the sparseness of formularized relationship and organized grammar. Somehow it reflects well the actual nature of the earth as it appears to the senses—prolific, confusing, rich, endless, abounding with possibility. Eventually it may become anachronistic, but until then, it is a treasure that the beginner must be made aware of and introduced to with care.

A Strange Stream

Geology encompasses the study of much that is dramatic. The shaking violence of earthquakes, the explosive eruption of volcanoes are physically dramatic events; questions such as the origin of the earth and the evolution of man are conceptually dramatic. But many of the events and questions that the geologist ponders are more subtle. They are the quieter niceties that constitute the body of the science. Their solution requires rigorous approach and provides deep satisfaction—no less than does contemplation of the larger, more dramatic problems. Indeed, consideration and understanding of the subtle is often the prelude to answers concerning the obvious. But beyond that, except on rare occasions, the subtle is what sets the tone of our lives and of the earth around us, and as such, invites our attention and appreciation.

In Wyoming and partly in Montana, there is a group of mountain ranges that surround a horseshoe-shaped valley that is more than a hundred and fifty miles long and perhaps not quite half as wide. The horseshoe, called the Bighorn Basin, is open on the north and leads to the high plains of eastern Montana; on the other sides, ringing it, rise the Pryor, Bighorn, Bridger, Owl Creek, Absaroka and Beartooth ranges.

The basin itself is dry. Clouds detach themselves continually from the snow-covered peaks to the west, pass overhead, and then disappear over the high rim to the east—but leave no rain. There are, in the summer, occasional brief storms, but the large drops that spatter and bead the dust evaporate before they can sink into the ground.

From the edge of the Beartooth, looking down into the horseshoe, the bare sediments of the basin appear as a pastel sea, undulating in the heat. Half way to the horizon, however, beyond the emptiness of the badlands, the Bighorn River, like the central vein of a leaf, winds down the axis of the valley; subsidiary veins branch laterally from it towards the mountains. You can't see the river itself, but you can see the irrigated strips of green-and-brown plaid that it feeds. Water in the Bighorn runs all year. The tributaries, fingering across the desert towards the basin walls, rush with life after spring rain and melting mountain snow; but otherwise, most of the time, they are quiet and dry.

The Bighorn River is the hand of God in this region, and it is a strange stream. It is not strange that it runs with water all the year, for it drains the inward facing slopes of all the surrounding mountains; every blizzard that lashes the eastern Beartooth, the western Bighorns, or the northern Owl Creeks seeps eventually, drop by drop through the basin sediments, into the channel of the river. It is also not strange that along its course, spaced every dozen miles or so, are a string of square, solidly prosperous towns that sit like jewels in the dust. What is strange is the way the Bighorn river enters and leaves the valley (Figure 1).

The river originates outside the basin. From its headwaters between the Owl Creek and Wind River mountains, it first flows east, parallel to the base of the horseshoe. Then, without hesitation, it makes a ninety-degree turn to the north towards ten-thousand-foot peaks—and cuts right through them. What a route to choose! Instead of seeking a path around the mountains, looking perhaps for a gap or a break, the river acts as if they weren't

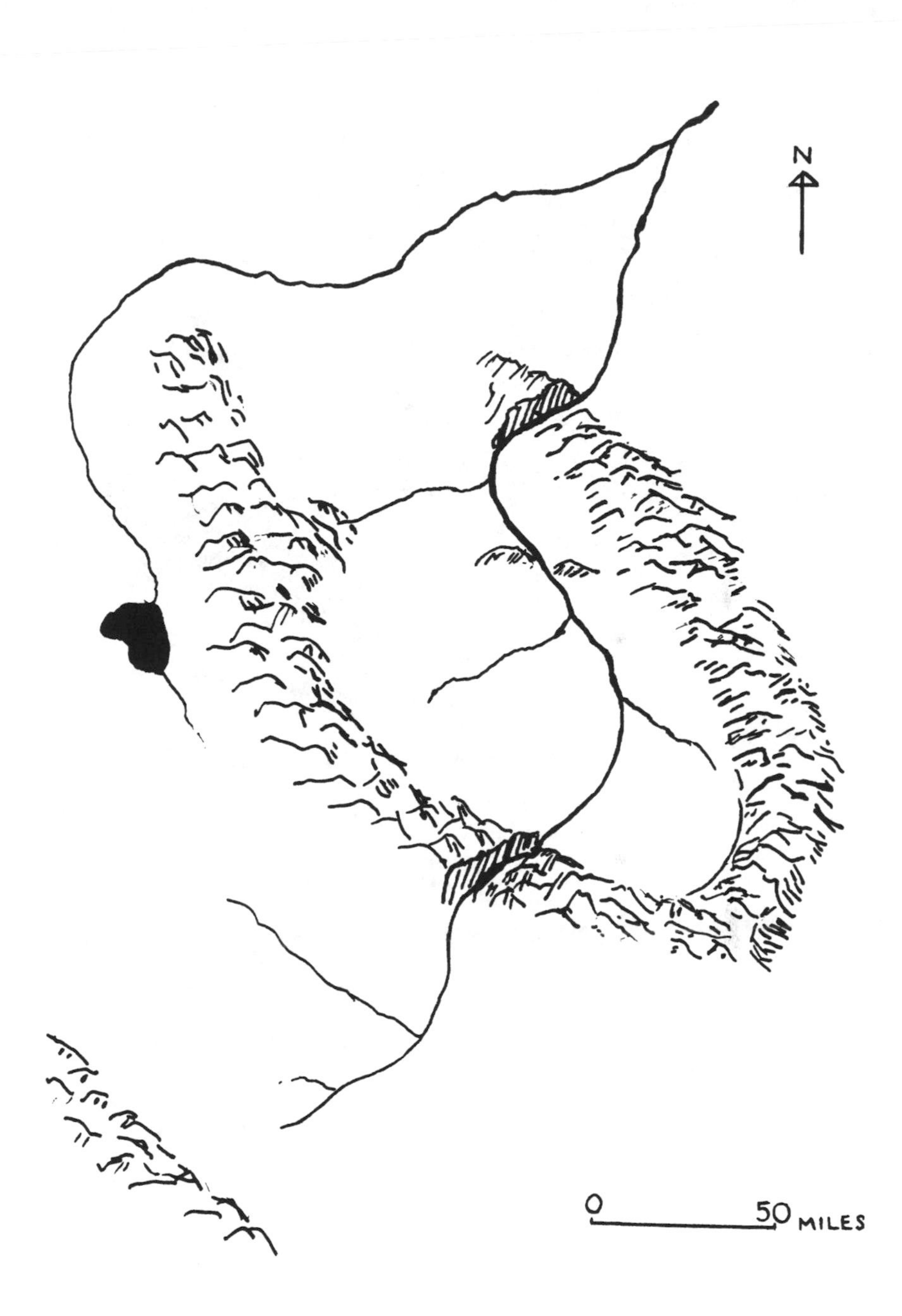

Fig. 1 What is strange is the way the Bighorn River enters and leaves the valley . . .

Fig. 2 It cuts right across the mountains through a spectacular chasm.

there. It runs through a gorge, steep-walled and dramatic, that cuts through to the core of the mountains and on out to the lowland of the basin floor on the other side. But the river is not content with this strange behavior. After flowing north along the valley floor, instead of leaving the horseshoe through the open, unmountainous end, it swerves at the last moment and cuts right across the northwest extremity of the Bighorn Mountains, through a spectacular chasm (Figure 2), to get out onto the Great Plains.

If you were to question the townspeople about this, they might wonder why you should think the river strange. They might point to the horizon, to the sky: God did it—he cracked the mountains and let the river flow through. After all, it has to go somewhere. It comes in through a canyon in the south; why shouldn't it go out one in the north?

And you might not argue, for not far to the west in the Absarokas, aren't there streams that flow hot, muds that bubble from the earth, steam jets that shoot to the sky? And atop the eastern ranges, isn't there a circle of rocky slabs as ancient and haunted as Stonehenge? There are still wild horses that run in the remoter parts of the basin; old men live in caves, guarding their secrets with guns. It is a primitive land; the ranches and oil wells are perched on the edge of the possible.

That an act of God or a catastrophic tearing of the earth opened the chasms in which the Bighorn River flows through the mountains seems more than likely—especially when the passive nature of most other streams is considered. In 1802, John Playfair illustrated James Hutton's theory of the earth in which he outlined the fashion in which rivers "hollow out" their valleys. Contrary to popular opinion at that time, he declared that rivers do not flow along preformed cracks or crevices; they excavate their own channels, form their own paths. Indeed, he said that if by "some sudden operation of nature" a mountain range were to rise and block the course of a river, the water, thus dammed, would rise until it found a low passage across or around the obstruction, which would then become its new path.

But the Bighorn River, flowing peacefully north towards first the Owl Creek and then the Bighorn ranges, does not seek an alternate route when it meets the mountains. Its direction is sure; it shows no signs of ever having considered the mountains an obstacle at all.

Indeed, if James Hutton had been a native of Wyoming rather than a Scot, he might never have arrived at the conclusion that streams carve their own course, for all over the Rocky Mountains major streams and numerous tributaries flow through high areas in deep gorges—rather than seeking a way around them. The challenge of these strange streams of the Wild West might well have delayed or frustrated Hutton's formulation of the Law of Uniformitarianism—that earth phenomena are the result of natural, usually gradual, observable processes. If he were to have watched the activity of the Bighorn River, he would have found no hint as to how it could have cut through the mountains. If he had placed a boulder in its path, like all other streams it would have made a detour around it. If he had dammed its gorge with clay, he would have seen it pond up behind the dam, rise, flow over it, and then cut a miniature canyon through the clay down to its former level. But if he were to have looked for evidence of such damming and ponding along the reaches of the Bighorn, or other such streams, he would have found none. The clue that leads to an answer, and we cannot dismiss the thought that a man of his perception might well have found it, lies far from the stream bed, most likely encountered only by chance wandering or systematic search.

To understand the implications of the clue, it is best to consider what does determine where most streams flow.

All streams flow away from high areas towards low ones, and then eventually to the sea or an inclosed inland lake or depression. The general initial direction of the route is determined, therefore, by the regional slope of the land. During and after rainfall, water flows downhill as a film or sheet, washing away loose or easily detached earth and rock. Since no land surface is exactly uniform, some parts will be more easily eroded than others, depending on how well consolidated the material is, what it is made of and whether it is fresh or deeply rotten and fractured. With successive rainfalls, the moving sheet of water will concentrate into rills and gullies excavated in the most easily removable material. With time the land surface is etched into ridges and valleys that reflect and emphasize the original variations that the running water encountered. Once the streams are well established and run along beds dug into soft material with high valley walls on either side, they are, in a sense, imprisoned. If, at depth, they encounter material that is resistant to erosion, they cannot escape it. They have no recourse but to slowly erode their way down through it. Only capricious acts on the part of nature can free the stream. For instance, landslides or lava flows may choke the valley, or the region may become tilted so that the stream eventually spills over its valley walls.

But what happened to the Bighorn River, to all the strange streams in the Rockies that cut through ridges? The clues that lead to an answer lie far away from the streams, high up on the sides of the basin walls. There, sticking like bits of plaster to the lower mountain slopes, are remnants of ancient stream deposits. Former streams, evidently, flowing off these same mountains, must have gradually filled the basins with sediment until the valley floors were considerably higher than now. In fact, the basin floors must have gradually risen so high as to ultimately connect, through low passages in the mountains, with the filling and rising floors of adjacent basins. Major streams, ancestral to the Bighorn and other modern rivers, were then free to meander across this high surface, passing at will between the diminishing peaks that still poked above the aprons of accumulating sediment.

Then something happened. The streams, instead of continuing to deposit, began to excavate, to downcut. The river channels were no longer able to wander across an essentially level plain, where each slight flood could alter their course, but were incised deeper and deeper until their routes were unalterable. With time the elevation of the high plain was increasingly reduced as the

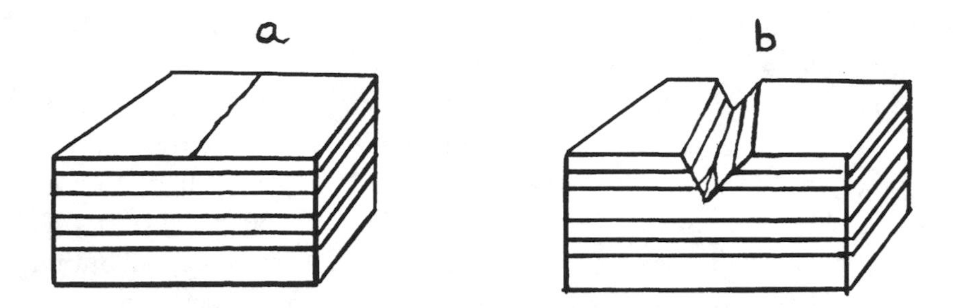

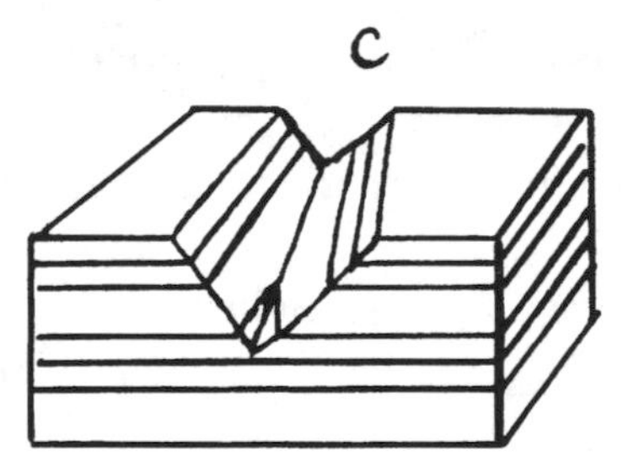

Fig. 3 A valley that has been cut gradually by the river that flows in it.

streams stripped off and carried towards the sea the sediment they had previously deposited. As the surface was lowered further and further the ancient buried ridges and basins were gradually exhumed. Where the vigorously downcutting waters were flowing transverse to the mountains beneath them, they had no recourse but to cut down through them too. Thus the routes of the modern streams were inherited from higher level ancestral streams independent of the buried topography they would encounter at depth.

There is, then, no sudden catastrophe required to explain the

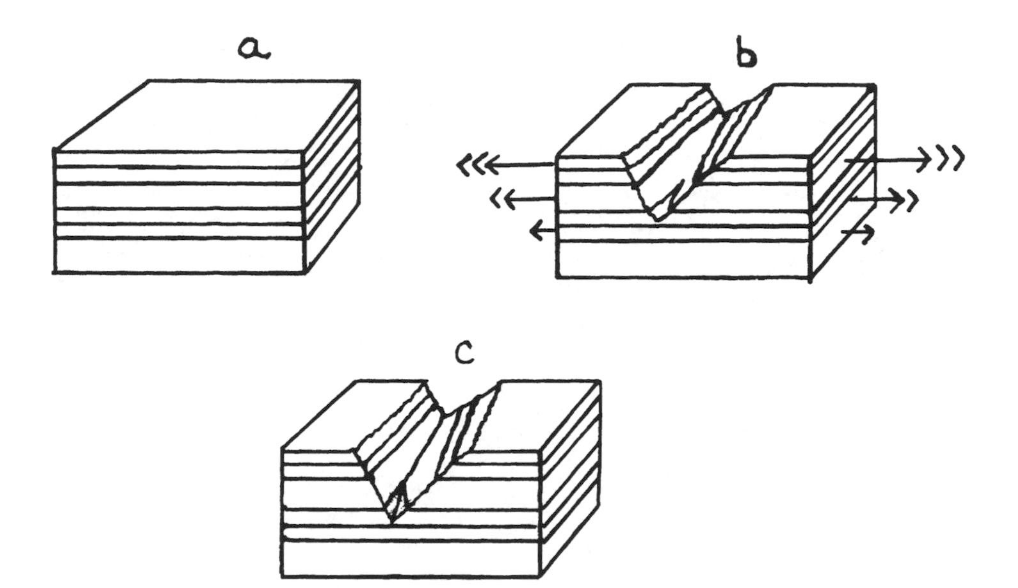

Fig. 4 A valley created by a "pulling apart" of the earth; a river then flows in the valley.

gorges through the mountains. The strange nature of the Bighorn River—its miracle—is the result of depositing streams becoming downcutting streams. Why this change took place is another question. Slight bowing up of parts of the region, if it served to increase the velocity of the streams, may have been the cause. The earth's crust constantly shifts, settles and buckles, so such a possibility is neither remote nor uncommon.

The result, though, is uncommon: potentially isolated arid basins gained water, life and connection with the outside world. Land that at most could have supported a few head of cattle lost its

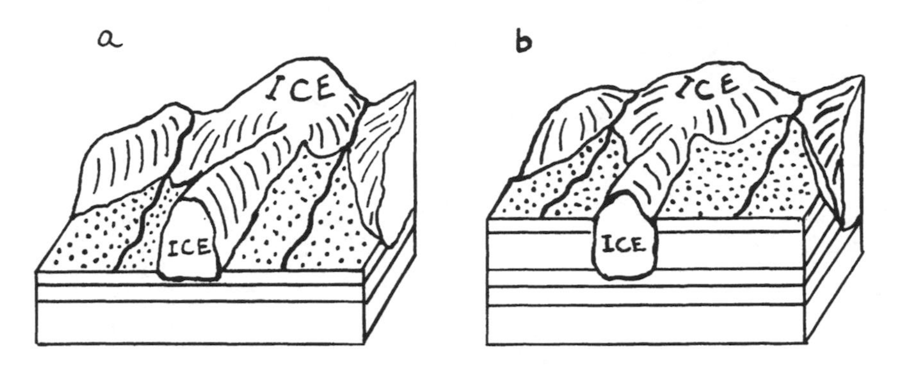

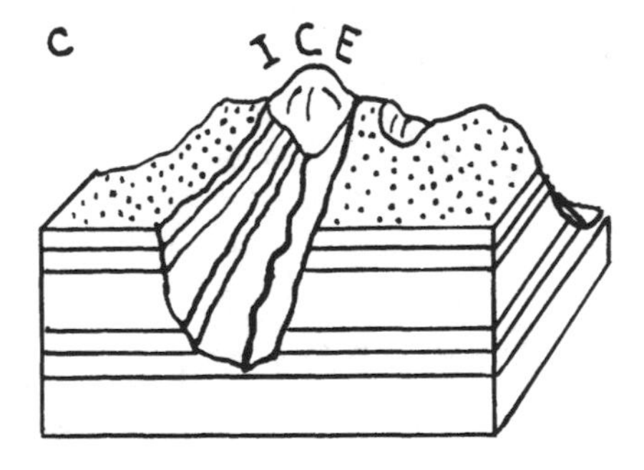

Fig. 5 A valley left when a sediment-covered block of ice melts; a river then flows in the valley.

loneliness. Even if the desert still laps across the Bighorn valley and the mountains rise implacably in every direction, there is a strip of oasis, a fragile accident of time, that runs down the center and permits the presence of man. Whether the sweetness of these coincidences is common knowledge to those who live there seems important. The prehistoric Indian dwellers raised monuments to the gods. Perhaps one day the stone circle in the mountains will find a grateful echo in the valley.

I Am Here—It Is There

A measure of whether life is meaningful or not is the extent to which it is involved in an investigation of the way things are. The "way things are" includes the way people are, the way the world is and, ultimately, the way "I" am. An artist, a scientist, a philosopher or psychologist, all can be considered part of the more or less organized phalanx that humanity throws into the quest for existential knowledge. But a psychologist can be obtuse, an artist jaded, a scientist misled. The overt activity at which a man spends his time has little correlation with his actual awareness. There is no doubt that the deadening nature of some occupations is a severe handicap, but the search for knowledge of the way things are is not the exclusive right or ability of any group. It is, rather, every man's domain and potential.

Children are all actively involved in discovery of the way things are. A not insignificant part of the investigation is the topology of their surroundings—the nature of the shapes and passages through which they move. Not content with passive examination of contiguous space and determined to see its relationship to themselves, they build: houses and castles out of books; tunnels and cities out of sheets and packing cases. They try out different realities, different arrangements, different dimensions. But as man grows older the materials change. We go to work, we raise families. We become enmeshed in the exigencies of "practicality," and all too often the flexibility is lost, the investigation abandoned.

But this is a mistake. To lose any of our faculties is devastating, even if not immediately painful. Rather than being abandoned with time, the investigation should be revived and expanded.

The consideration of where we are is an important part of the understanding of the way things are. Whether at any moment we are at the top of a mountain or walking across a plain does make a difference. The physical and metaphysical are ultimately inseparable. Constant rebirth, rejuvenation, without which we can only slip towards death, requires incessant pilgrimage—the passage from here to there, the journey more important than the goal. If we can't move physically, at least we can move in our minds. It is the wandering that counts, the moving along a path to reality, a route to ourselves.

One common result of travel is maps. Symbolically, actually, they mark where we have been and where we may go. In the old days, the task of reducing the three-dimensional world to a two-dimensional analogue was an art. The surveyor (usually a geologist or military engineer) surveyed, determined exact elevations and positions of key points, and placed this information onto a base sheet in the form of numbered pinpricks. When all the necessary measurements were assembled before him, he then integrated and transmuted the seemingly scattered data into contours: flowing lines whose spacing and curves contained—to the trained eye—the essence of the land. This last act of creation was, as it should be, done in the field, with reality at hand. The result—a topographic map—a graphic summary of elevation, relief, abruptness, angularity, dissection, inclination. Today, the same work is accomplished by technicians in an office who peer through optical devices at aerial photographs. Efficiency has been gained; contact with the land, with the way things are, has been lost.

That this is a loss might or might not have been appreciated by the early giants of the topographic surveys. But on one point they would have agreed: land that is not mapped is not possessed. Expert tracing of contours is not unlike love-making—the earth becomes known—and the enterprise of the old-style topographers was in every sense an embrace of the planet. They climbed the peaks, scaled the cliffs, crossed the deserts. What they saw they distilled and transmitted as maps. What they felt we can only imagine—but it was perhaps more satisfying than the relief of a coffee break or a trip to the water cooler.

If the character of map making has changed, the nature and potential of maps has not. Their uses, of course, continually increase and diversify. For the geologist, topographic maps—those that show the shape of the land—are the frame to which the flesh of geologic knowledge may be applied: the sequence and distribution of formations, the inference and fact of economically valuable deposits, the occurrence and likelihood of water—in short, the information found on geologic maps. Pursue the argument, and topographic knowledge is preliminary to speculation concerning the origin of landforms, continents, the earth itself.

But beyond all these, in a way, prosaic, practical and scientific uses, there is an echo within maps of the reality they represent. Consider a map that is bare of all scientific, military or political symbolism, a map that is just a statement of configuration. In itself, doesn't it have much to recommend it? Even maps of boundaries, the shorelines of kingdoms—real or imagined—have possibilities far beyond where France ends and Belgium begins. They are a *tabla rasa*, a stage ready for creation or re-creation, the stuff of dreams.

Indeed, there is a power about maps. With a blank piece of paper and a sharp pencil you can create the Himalayas and then climb over them, form a swamp, shape an island. With a map before you, you can ask, "How would my life be if I lived there? Or here? Is there peace in that valley? How does the wind blow along that ridge, up that slope?" Even "Where would I place my neighbor? The nearest town? The city?"

Is this escapism? Quite the contrary. It is contact; advance rather than retreat. Like painting or sculpture, it is an arranging of shapes, forms in space—but with the proviso that the artist place himself within the arrangement. The creator becomes one of the elements of his creation, a mobile element, to be sure, one that explores the essence of the rest.

This kind of thinking is not original; it is an extension of the approach of the architect, but goes beyond structures physically created by man. It is an attempt to relate man to the universe, to keep him aware of the ultimate frame of reference. Through it, man can know where he is and what he is. The use of maps in this sense—as a tool—is only one way among many. But the interaction between man and his environment—in the most literal sense of "I am here—it is there"—has been neglected. The neglect is to the detriment of both. Try maps as an approach to the way things are.

Mapping

Mapping, *geologic* mapping, is a unique aspect of geology. To my mind, more than anything else it distinguishes and typifies the science. In order to *map* geologically you must be able to *see* geologically, and if you can't do that, the measure of your success is severely limited.

Geologic mapping, like other mapping, is a way of converting complex, abstract data into a visible form that not only conveys information, but also its implications. Consider a political map, for instance. Such a map shows boundaries between political units. But often the drawing up of such a map is more than a statement of fact—it may be the giving of an opinion, a declaration of intent or, indeed, a political act. Those familiar with the mechanics of local government know that when a proposed highway, park or school has found favor with the appropriate planning commission, it is "put on the map" and is well on its way from idea to concrete reality. Similarly, the creation of a geologic map embodies the summation of the working philosophy of the geologists involved. Its bright colors and clean lines may be the result of months or years of physical struggle and professional controversy.

As I mentioned earlier, a geologic map is a map that purports to show the distribution of geologic units at the surface of the earth. How these units are defined and what is the nature of their distribution is where the problem lies.

It would seem reasonable that before a map is begun, what is going to be shown on it should be pretty clearly in mind. A topographic map, for instance, attempts to portray the three-dimensional configuration of a land surface. Conventionally, such maps include contour lines, elevation marks, drainage, marshes, swamps, sand deposits, roads, buildings, section, county, state and international boundary lines, and latitude and longitude. The surveyors know the nature of what they will end up with at the start of their endeavor, and the creation of the map is pretty much routine application of well-established techniques. Geologic mapping in some cases is almost as straightforward. Often, however, the geologist may have no idea of what the final result will be, and the techniques he uses and his concept of what he is aiming for must change and develop as he proceeds.

The information the geologist wishes to assemble and place on a map depends first of all on what his field of special interest is. If he is interested, for instance, in *surficial* geology, the map will show the distribution of loose materials such as river sediment, glacial debris, beach deposits, sand dunes and so on. The assumption made is that he can distinguish between these different materials and infer their presence where they are covered by vegetation or man-made structures such as roads and towns.

If he is interested in *bedrock* geology, the map will attempt to show the distribution of consolidated igneous, metamorphic or sedimentary rock units that underlie the loose, surficial material. Regions where outcrops of bedrock are abundant include recently glaciated areas where moving glacial ice has scraped away most loose surface material, floors of deserts that have been swept clean by the wind, and coastlines where wave action has been particularly erosive. In most other places, bedrock is covered by vegetation or loose sediment except, perhaps on the sides of steep cliffs, in the beds of swift streams, or where roads, mines or quarries have been cut or blasted through rock.

Thus, for both types of study—surficial or bedrock—one of the initial problems is the degree of availability of material for examination. In both cases the unknown must be inferred from the known. Usually, the problem is most difficult for the bedrock geologist since the likelihood of cover is greater.

Once the geologist has decided upon an area to investigate he must come to a decision as to how to subdivide the rocks he is working with into mappable units. That is, he must decide upon what he considers to be significant differences between the rocks and how to draw boundary lines between them. If, for example, the region in question is underlain by shale, limestone and sandstone, he may decide that each of these rock types constitutes a mappable unit or "formation." If the region is underlain entirely by one rock type, say limestone, he may subdivide it on a more subtle basis, such as fossil content or reactivity to hydrochloric acid. Part of his decision depends upon how thick the individual rock-type layers are; if some of the layers are extremely thin, he may have to group them together in order for them to be visible at the scale he has in mind for the map. In turn, the scale he uses for the map depends upon the detail he needs in order to show what he considers is of interest.

In principle, the choice and recognition of units may sound simple. Problems arise, however, if the units fail to maintain their identifying characteristics over the distance separating outcrops. Such changes in character are not uncommon. Rocks formed in a sedimentary environment reflect local conditions under which accumulation and lithification of sediment took place. A brief glance at places where sedimentation is taking place today reveals lateral and vertical variations are often the rule rather than the exception. Consider, for instance, the sediment deposited where a river enters a large body of relatively still-standing water, such as the sea. It may be that on most days of the year the river water transports only a mixture of mud and silt; when the river reaches the end of its journey its velocity and capacity for carrying sediment are drastically reduced: the silt is released and settles to the bottom of the sea quite near the shore; the mud, being finer, is carried somewhat further from shore before it loses velocity and settles. Thus, in this situation, normal sedimentation produces a lateral variation from coarse to fine material going away from the shore line towards deep water. During rainy periods, however, the velocity of the stream may increase dramatically, and the water may sweep sand, pebbles or even boulders down to the sea. When this mixture reaches open water it deposits with the coarsest material settling near to shore and the finer further away. The pebbles and sand may settle where silt normally accumulates; the silt may be swept far out and deposit where only mud settles during drier seasons of the year.

The very nature of this sedimentation process, therefore, makes vertical and lateral sediment variation inevitable, and this variation is preserved in rock when the sediment is lithified. Furthermore, variation in sedimentation conditions are accompanied by variations in types of life existing under these conditions. In the case above, zones of distinctly different life communities parallel the shore line of the sea. Fossil evidence of this life preserved in rock reflects these variations.

Thus, the definition and recognition of units for mapping is often complex. Indeed, it may be the purpose of the investigation to discover, trace and understand these variations.

Where sedimentary materials accumulate without significant local lateral and vertical variation, units may be easily defined and recognized from outcrop to outcrop for tens or even hundreds of miles.

When the geologist has come to terms with the problem of map units, his next task is to discover their configuration: the shape, extent and nature of their boundaries. Most sediments accumulate as layers or "beds" that pile up like a series of blankets on the floor of an ocean, the bottom of a lake, or the floodplain of a river. The surfaces that separate the blankets from each other mark halts in deposition or abrupt changes in the type of material being deposited. These surfaces are referred to as "bedding planes" and are highly characteristic of sediments and sedimentary rocks. Indeed, one of the criteria for recognizing sedimentary rocks is their layered nature. When formed the layering is usually horizontal. If the rocks have been disturbed subsequent to their formation, the layering may be broken or thrown into a series of folds that are a record of the deformation. Usually the geologist cannot see a complete fold at any one outcrop. He has to infer its presence and character from the attitude of tilted bedding planes at scattered outcrops, a problem somewhat akin to visualizing the picture on a jigsaw puzzle when most of the pieces are missing.

If the rocks have merely been folded, the geologist's task is comparatively simple, for once he understands the style of the folding he can put together a reasonable reconstruction of the whole from examination of just a few parts. If the rocks have been broken and shifted as well as folded, the problem is more formidable. Material that was originally adjacent may have become separated by anywhere from a few feet to a few miles. Not only that, rocks that have been broken and shifted in this manner, or *faulted* as the geologist would say, may, at a later date, have been faulted a second or even a third time. When this has happened, the originally continuously layered rocks now have the appearance of a haphazard mosaic. In such a case, inferring the relationships between patches of rock in separated outcrops requires considerable skill and ingenuity.

Another complication arises when layered rocks have been tilted or folded so that in places the sequence of layers is inverted. If at an outcrop the map units are easily recognizable, this presents no problem; experience in adjacent areas will tell the mapper which layer should lie on top of the other. But if the identity of the rocks is ambivalent, a wrong guess as to whether the sequence is right side up or inverted can result in a reverse picture of the nature of the folds: upwarps will be taken for downwarps, folds

(Overleaf) Arches National Monument, Utah

that die out to the east will be thought to die out to the west. The solution to this problem lies in the recognition and discovery in the rocks of features that have right side up and upside down unmistakably built into them. For instance, when ripple marks form in sand on the floor of the ocean, their troughs are rounded and their crests are pointed. If these ripple marks are preserved in rock, the points of the crests will indicate which direction was originally up. Through careful observation of modern sedimentation, lists of hundreds of such telltale features have been compiled. Often, however, none of them may be present at a key outcrop, and other less-certain solutions must be sought.

The geologist engaged in mapping, therefore, must have a wide background as to the nature of geologic materials and processes as well as a mind that can extrapolate scattered two-dimensional information into a coherent three-dimensional picture. Mapping, in turn, aids in the understanding of materials and processes. It permits meaningful assembly of fragmentary data and serves as a vehicle to express working hypotheses.

As the map develops, it records the geologist's decisions as to what units are present at each outcrop, the attitude of their layering and other structural features. Furthermore, as he proceeds his inferences as to what lies between outcrops can be put on the map —in dotted or solid lines according to the depth of his conviction. Most important of all, as the pattern emerges the map reveals where information is scanty and where outcrops, if they are present, might reveal the crucial nature of a fold or a fault; in short, it tells the geologist what he must do next.

Problems of mapping sediment, loose or consolidated, are simple compared to those encountered in the investigation of metamorphic rock. In fact, until quite recently, complex metamorphic rocks were considered unmappable. They incorporate all the problems of sedimentary rocks plus additional complications born of metamorphic processes.

The mineralogy, texture and chemical composition of any rock is controlled to a large extent by the environment in which it formed; if the environment of the rock is changed subsequent to its formation—more specifically, if the temperature, pressure and chemistry of its surroundings change—then the materials present in the rock may alter their size, shape, composition and distribution accordingly. Such a process is called metamorphism.

A rock that has been metamorphosed may well lose all those characteristics that served to identify it in its premetamorphosed condition. A sedimentary rock, for instance, may be greatly thickened in some places, stretched and thinned in others, pulled apart so that its continuity is destroyed, have its layering obscured by superimposed metamorphically developed layering, change composition radically both laterally and vertically, and be intensely folded. In addition, it may undergo such transformations two or three times in response to several successive periods of environmental change. Also, such change may be accompanied by invasions of molten rock that will produce further disruption and chemical alteration. Lastly, the material in question may itself locally attain temperatures sufficient to melt, and thus blot out even its metamorphic characteristics and become, when it resolidifies, rock that *must* by definition be called igneous.

The geologist investigating metamorphic terrain seeks to unravel the details of such events. He wants to know the conditions under which each period of metamorphism took place, when they occurred and what were the specific responses of the rock to its changing environment. Ultimately, he would like to know the nature of the original, premetamorphic rock and its conditions of formation. In short, he wants to trace the geologic history of the region. To do all this mapping is essential. It is the only way to assemble the information. Before he can understand *how* and *when* something has happened, he has to know *what* has happened.

His first job is to choose a base map on which to plot his field data. Just as in the mapping of sedimentary formations, the spatial relationships between information gathered at each outcrop of metamorphic rock is extremely important. If the map is to be very generalized, he could use a standard one inch or one-half inch to the mile topographic map, such as those issued by the government. If he needs more detail, he may use areal photographs of the region. Areal photographs, however, are of limited use in heavily wooded areas, for it is almost impossible for the geologist to locate himself and know where to plot the data. In such cases, or where he needs extremely fine detail, he must create his own base map. That is, he must employ surveying equipment to construct a network of known points from which to work.

But it is at the outcrop itself that the geologist's true skill is put to the test. It is here that he succeeds or fails. The prime prerequisite is that he be able to *see*. Facing him are an array of lines, layers, and textures that must be sorted out, the important distinguished from the unimportant, the accidental separated from the relevant. If he makes an error or an incorrect assumption, the map will be wrong or, more likely, inconclusive.

When a non-geologist looks at an outcrop his attention may be drawn to its overall shape, its general color and, perhaps, whether or not it seems a likely place to sit down. All of these aspects to a large extent are accidents of erosion and weathering. These processes do somewhat reflect the structure, texture and mineralogy of the rock, but on the whole the geologist must envision the rock as if it were fresh and uneroded. Before him, perhaps, is a metamorphic rock whose surface has been deeply grooved by glacial gouging. These grooves may cut across one or more earlier sets of parallel fractures (or *joints*) that were the rock's response to the last regional deformation. The fractures may have been subsequently filled with a resistant material such as quartz, in which case they will stand up as minute ridges, or they may form linear recesses if they were filled with a substance more soluble than the adjacent rock. Both fractures and grooves, however, are relatively recent superficial patterns superimposed upon more fundamental structures within the body of the rock itself. The rock may display a metamorphic layering: layering produced in response to intense stress, temperature and chemical gradients present during an interval of its history when it was fairly deep beneath the surface of the earth. Such metamorphic layering may be an intensification of earlier sedimentary layering, or may be quite independent of it. During or after formation, the metamorphic layering may have been folded—and then folded again—and then folded once more. Each period of folding may create new sets of metamorphic layering that partially or totally obliterate earlier formed ones, and traces of still earlier, original premetamorphic sedimentary layering become increasingly or impossibly obscure.

An additional complication arises due to the fact that the very act of deforming a rock often results in a redistribution of its chemical and mineral constituents; in turn, the distribution of these constituents controls, in part, how the rock will respond to deformation.

The geologist's task is a tough one. He is, in a sense, caught in a vicious circle. In order to sort out the mess that faces him, he must have some idea of what has happened. If he has no concept of what processes have been involved and how many times they have operated, the likelihood of his integrating the lines and layers present in the rock into a meaningful pattern is remote. On the other hand, to understand the processes requires knowledge of what is before him: the distribution of rock types and the geometry of their structures, knowledge gained only from knowing how to plot what he sees on a map.

What happens in actuality is that the geologist must somewhat arbitrarily make some initial assumptions in order to proceed. If the assumptions produce meaningful results, fine; if they don't, and this may be the case time and time again, he must start over, make different assumptions and see if the results are better.

This, perhaps, is just normal application of the "scientific method." There is an inherent danger, however, in this procedure. In all likelihood the assumptions he makes are based upon his previous experience—experience in the field or that gained from studying other people's work. It may be, however, that none of this experience is completely relevant to the problem he is working on, and in making assumptions as to what processes have operated, he may be, in a sense, negating the possibility of discovering new processes. This is most likely when false assumptions produce feasible results—a map that explains *almost* everything the geologist sees. Now, probably no geologic map ever published represents one hundred percent certainty on the part of its compiler. Nor should it have to, otherwise few maps would ever appear in print. But what may happen is that as the geologist sees his assumptions apparently working, he may discard anomalous information as unimportant—which it may or may not be. After a while he may not even see such features in the rocks anymore. (If thy right hand offend thee . . . !) Thus the very act of seeing a pattern emerge in the rocks causes one, consciously or subconsciously, to disregard elements that interfere or clash with this pattern.

This problem, I am sure, is not unique to geology. It rests in the overwhelming human urge to find order where there appears to be only chaos. Science, in other words, no matter how hard it strives towards and loudly claims objectivity, ultimately and inevitably is subjective: its foundation stands on the unprovable assumptions of Uniformitarianism; the choice of which evidence to consider is of necessity arbitrary and personal.

Understanding and mapping metamorphic rocks, therefore, is like trying to follow a novel in which the characters constantly change name, sex and occupation, and whose plot and style alter constantly in response to the feeling and attitude of the reader. The author's original intent is not only obscure, but secondary; the act of reading is as much creation as the act of writing.

Similarly, a geologic map is just as much a statement of the geologist's theories and opinions as it is of geologic fact. J. M. Harrison, in an article entitled "Nature and Significance of Geological Maps," has made the point nicely. He puts side by side two maps of the same area, one mapped in 1928, the other in 1958. To the inexperienced eye they would have little in common. The number of units, the names of the rock types, their distribution and configuration, are all different. The maps reflect the changes in

Sage Creek, near Interior, South Dakota

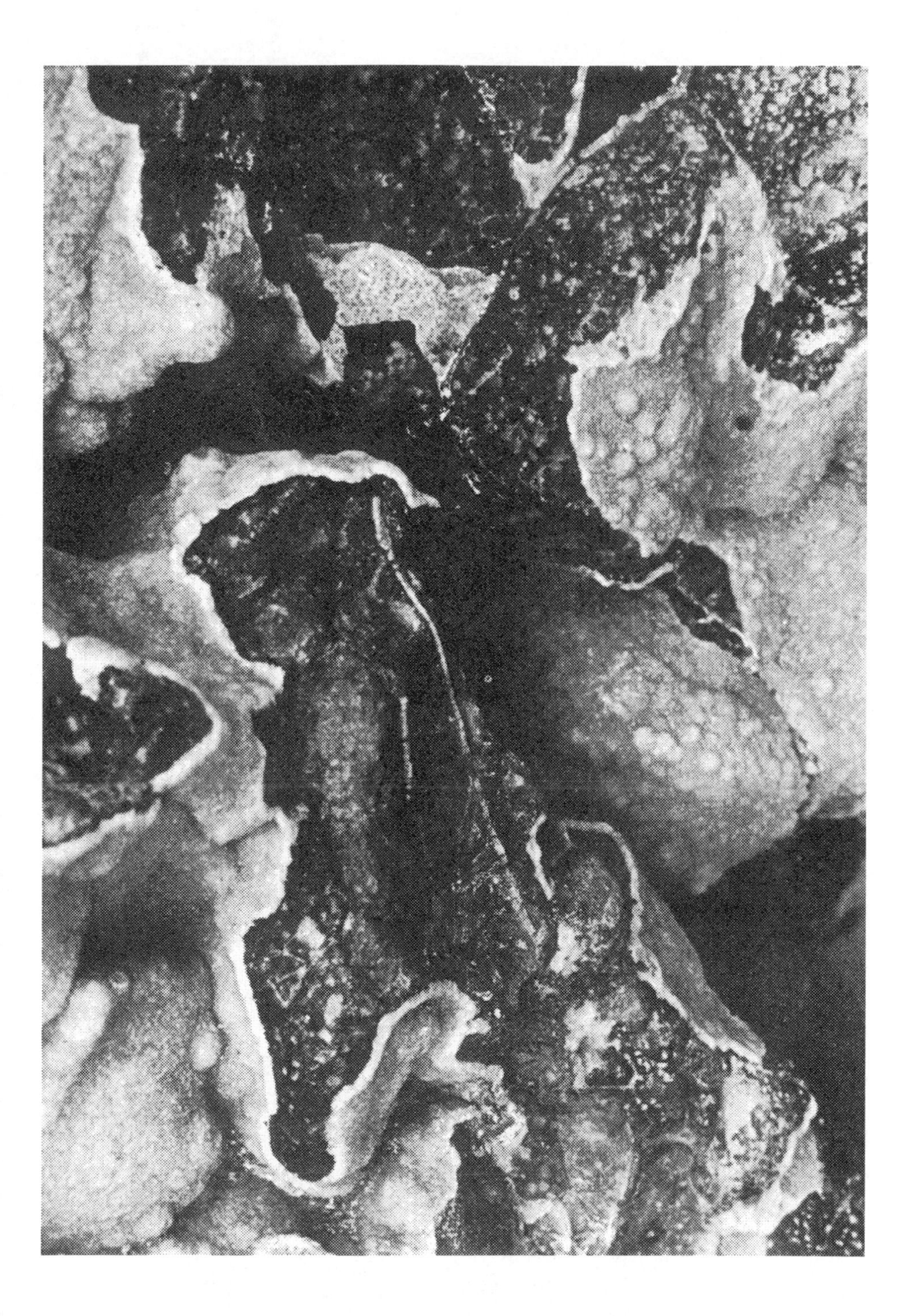

geologic thinking that took place over thirty years. The rocks didn't change. What the geologists saw did. It's almost as if each outcrop was a geological Rorschach test. Other geologists must decide whether the mapper's interpretations coincide with their own prejudices. Maps are not really a question of truth, they are what emerges after geologic measurements pass through the filter of idiosyncratic personality.

Geologists often claim that geology is as much art as science; if maps are the essence of what is geological, then they are probably quite right.

Seeing the Unseeable

The legitimate domain of geology includes the earth—its interior, its surface, its seas, its atmosphere—and all that has happened to it during its five billion years of existence. It is, however, one thing to claim a kingdom, quite another to rule it. For all the glory of the geologic realm, most of it lies beyond the reach of the geologist: obscured by the mists of the past, covered by the depths of the oceans, buried under tens, hundreds or thousands of miles of rock. The geologist runs around on the small patches of the earth available to him and thinks about the rest. It is not a gloomy picture, but it is a situation that demands the utmost ingenuity—practical and theoretical.

There may in the future be a time when all the earth is available for direct or instrumented observation. Perhaps H. G. Wells's dream of a time machine will one day be realized and permit the evolution of the planet to be watched—preferably from a comfortable armchair. But until then ideas must depend upon fertile imagination and sound logic. The unknown must be approached from the known. Each scrap of information must be examined for its implications—especially in fields seemingly unrelated to it. Sometimes the results are extraordinary.

Consider, for instance, the way in which the question of when the moon first became associated with the earth has been approached. Preston E. Cloud, Jr., in an article entitled "Atmospheric and Hydrospheric Evolution on the Primitive Earth," states that the shape of certain sedimentary structures built by marine algae—known as stromatolites—are particularly relevant. Stromatolites seem to have been around for a long time. Remains of these structures are found in rocks that are more than two billion years old, and they are still forming in certain regions today. When stromatolites form in intertidal environments at the ocean's edge, they may attain a vertical height or relief of up to two feet. Those that form in regions of lesser tidal range, however, are typically low-relief crusts and nodules. Cloud believes that the existence of high-relief stromatolitic algal domes in ancient rocks indicates the existence of substantial tides at that time. Such tides, he concludes, could not have been due to the action of the sun alone, but must have been generated principally by the gravitational influence of the moon. The moon, therefore, he concludes, has been in close association with the earth for at least two billion or so years.

If Professor Cloud's approach to the age of the earth-moon relationship seems tenuous, it is not unique. In a similar vein, Lamar and Merifield consider the relationship between "Cambrian Fossils and Origin of Earth-Moon System." Their conclusions, however, differ somewhat from Cloud's.

One of the consequences of the earth's association with the moon is a gradual increase in the length of the day: tidal action produced by the moon (combined, of course, with that produced by the sun—the strengths of their respective effects are in the ratio of 11:5) acts as a brake and slows down the rotation of the earth about its axis. Revolution of the earth about the sun, however, is not affected by this braking action, and even though the days are gradually lengthening, the time encompassed by a year remains the same. Thus, the *number* of days in the year decreases as the length of each day increases. From astronomic and geophysical calculations, it is possible to arrive at an estimate of when this tidal braking action began. The answer is 0.5 to 1.8 billion years ago. The upper figure is less than that given by Cloud, the lower figure is considerably less.

The next step in Lamar and Merifield's argument involves fossils of corals. Corals are relatively simple animals that live attached to limy structures of their own making. These structures are built and enlarged as the animal grows. Daily growth is manifest on the structures as a series of lines—similar to the seasonal growth marks preserved as rings on cross-cut tree stumps. Study of corals throughout the geologic record reveals that the number of daily growth lines per year (each year being represented by a cyclic change in the thickness of growth lines) has decreased from about four hundred in the Devonian (about three hundred and eighty million years ago) to an average of 360 at the present. The rate of increase in the length of the day suggested by these figures agrees with that afforded by astronomic observation and supports the 0.5 to 1.8 billion year age for the start of the earth-moon relationship. Indeed, they say, the figure suggested by growth lines is 0.7 billion years.

In order to continue the story, a slight digression is necessary about the nature of the fossil record. Fossils are non-existent or extremely rare in rocks formed during the first four and a half billion years of the earth's history. The algal remains mentioned before are an unusual exception. The absence of fossils does not mean the absence of life. The ways in which preservation is accomplished favors creatures with hard parts such as shells or bones. Early life forms may have had neither; they may have consisted of jellyfish, worms and so on, species that lacked a sense of responsibility to history. At the beginning of the Cambrian, however, about six hundred million years ago, abundant life forms with hard parts capable of preservation suddenly came into being, as indicated by their abrupt appearance in the rock record. This life was abundant not only in number of individuals, but in diversity of forms.

How did this happen and what did it mean? Lamar and Merifield suggest it was the result of the capture of the moon and the consequent increase in oceanic tides. Increased tidal activity and the resultant turbulence in shallow water made the possession of a shell a distinct evolutionary advantage. The catastrophic events associated with the moon's capture may also have paved the way for the development of creatures with hard parts by causing the extinction of many species without hard parts, thus opening up many ecologic niches into which surviving species could radiate. If this speculation is correct, then the moon has been companion to the earth for only six or seven hundred million years —as also suggested by the study of coral growth lines.

Whether Cloud is correct, or Lamar and Merifield are correct, time, perhaps, will tell. It is interesting to note that in the last argument fossils give information about the moon, but the moon also suggests ways of interpreting fossils. One hand washes the other. Paleontology and astronomy join together to examine a specific event in the past.

The interior of our own planet is, in a sense, more remote than the moon. Space travel has begun; journeys to the center of the earth have not, nor do they seem imminent. The major structure of the earth, as far as we know, consists of a thin, outer crust, a region of intermediate depth called the "mantle," and a central core. The crust is all we know; it is what we walk about upon. The mantle and the core we have never seen. Study of earthquake waves, geomagnetism, gravity variations, together with heat-flow measurements, more accurate knowledge of the shape of the earth, and consideration of how the earth and moon affect each other all help put certain restrictions on what the interior regions must be made of, but they do not put a chunk of it in our hands. Some material that may have originated within the mantle is found as inclusions within deep-seated igneous rocks or "floating" within lava erupted from oceanic volcanoes. But, pardoxically, it is through examination of material from outer space that we may make some of our best guesses about what's under our feet.

According to Bode's Law of planetary spacing, a major body should be found between Mars and Jupiter. Instead of a planet, however, there is a swarm of meteoric material that forms what is known as the "asteroid belt." Many of the chunks of material that reach the earth from outer space, the so-called shooting stars or meteorites, probably were members of this belt that had exceptionally eccentric orbits and came close enough to the earth to be captured by its gravity. These meteorites are of two distinct types: metallic and stony. The former are iron-nickel alloys; the latter are composed mainly of the mineral olivine—a magnesium iron silicate. The presence of olivine, and its occasional association with other minerals such as diamond, has been taken to indicate that such meteorites formed in a high-pressure environment, one that might be expected in the interior of a large planet—one perhaps of earth-size. If this were so, maybe the asteroid belt is the remains of such a planet that once existed between Mars and Jupiter and that has, for some reason, shattered. If this hypothetical planet existed and was of earth size, it may, like the earth, have had a crust, mantle and core. If so, the dense, metallic meteorites might be remnants of its core; the less dense, stony meteorites remnants of its mantle. Continuing the analogy, these substances then become likely candidates for what makes up the core and mantle of the earth. Nickel-iron alloy, as found in the metallic meteorites, meets chemical and physical requirements for core material, and furthermore, helps explain the origin of the earth's magnetic field. The olivine-rich stony meteorites are similar in many ways to the "inclusions" brought up from great depths by magma and lava, and they or some equivalent may be what makes up the mantle.

In both examples that I have cited—the question of the age of the earth-moon relationship and what is the nature of the earth's interior—direct observation is not possible; tentative answers, however, have been arrived at through elaborate chains of reasoning based on what, at first glance, seems irrelevant information. But whenever a conclusion is so far removed from basic data, its reliability should be examined closely. If any of the preliminary

or intermediate assumptions should prove false, the whole edifice is in danger of collapse. The greater danger, though, is that the edifice may not collapse even if its foundation should prove untrue. When chains of reasoning crisscross back and forth over the boundaries between scientific disciplines, difficulties in flow of information and lack of background on the part of investigators may hinder their understanding all the implications of the data they deal with.

Wasserburg, Sanz and Bence have recently suggested that metallic meteorites need not have formed in a planetary core. Moreover, Brian Mason, an important worker in the field, thinks that the minerals present in meteorites indicate that meteorites stemmed from the breakup of not one large, earth-size planet, but from the destruction of many small planetoids, none of which was more than about one twentieth of the radius of the earth.

If true, this information certainly weakens some of the links in the chain of reasoning outlined earlier as to what is in the interior of the earth. Whether the chain must break is not clear. It is quite possible that while the reasoning may change, the conclusion will remain the same.

One last comment seems appropriate regarding information sources for unobservable phenomena. To be eclectic pays off scientifically. No field is really trivial or unimportant; no field can be considered unalterably obscure. Not so long ago the phenomenon of radioactivity was considered interesting but outside the mainstream of scientific thought—a subject worthy only of idle curiosity. The passage of relatively few years, however, has turned this "curiosity" into both the hope and despair of mankind. Radioactivity on the one hand may be the energy that ensures a future of peace, health and prosperity, or on the other it may be used to cause the catastrophe that denies man any future at all.

Detail, Goosenecks of the San Juan River, Utah

The Relatively Absolute

A strange phenomenon of our time is that we seem to be impressed by our own insignificance. The larger the universe, the more remote our celestial neighbors, the smaller the earth in comparison to giant planets and stars, the better. The same holds true for the temporal dimension. We are awed by the geologist's assurance that the age of the earth is measurable only in years that number in the billions, or that a given rock has endured for a healthy fraction thereof. This inclination for belittlement is a recent development. From the Middle Ages to the start of this century, the trend was quite the opposite. Everything possible was done to reassure man that he was the center of a universe designed by God for him. Furthermore, the duration of this Divine whim was at most a few thousands of years; that is, long enough only to accomplish the unraveling of biblical events.

But both of these seemingly contrary emotions, the delight in insignificance and the urge to be the focus of the universe and of God's attention, have in common an interest in and an awareness of man's place in space and time. Man may be short-lived and of infinitesimal size, yet to the extent that he is natural, he shares in the glory of nature's magnitude and, moreover, whatever little he does accomplish is that much more impressive in the face of such vastness. Or, alternatively, if man is the center and end point of things, then his every action and decision is of moment and bears the weight of great responsibility—in terms of the universe that serves him and in terms of God whose servant he is. Furthermore, man of today and man of the Middle Ages are both endowed with an intense historical sense—a sense of coming *from* somewhere and of going *towards* somewhere. That is, they are both *Western men*, stemming from and developments of classical Greece and Rome and possessed of an ingrained faith in cause and effect.

But what about other men, men alien or who lived prior to the origin of Western man? Thomas Mann, in the Prelude to *Joseph and His Brothers*, presents an unforgettable image of the early Palestinian living in an unending and undifferentiated present. For man living then, time had no direction, no historical sense. Past, present and future were vague distinctions. It may have been that the flight of Joseph to Egypt and his triumphal return were events that happened only once to one person. Or it may have been that the events happened only once, but the Joseph who went to Egypt was not the same Joseph who returned. Or it may have been that both the events and the person were archetypal, that the saga had already occurred over and over again throughout the mists of time, was occurring at that particular moment, and would occur again and again, without end, as long as man existed. Which of these possibilities was "true" was both unclear and unimportant.

How can we, today, conceive of a world that is not historical? According to Oswald Spengler in his *Decline of the West*, the man of Classical Greece (perhaps like Mann's early Palestinians), *lived in the present* and lacked Western man's historical sense. Furthermore, this lack of historical sense was not a superficial difference separating Western from pre-Western man but was so fundamental as to make it impossible for a man of the one culture to envision the universe as it appeared to man of the other culture.

Thus modern man, as the most recent manifestation of Western man, is left inescapably involved, enmeshed and fascinated with time. But what it is, he doesn't know. The study of the true nature of time has become the province of the mathematical physicist. It is, we are told, not really different from the spatial dimensions: length, breadth and height. It can only be measured and, therefore perhaps, only exists where there is change. If there were no change, no movement, how could time and its passage be conceived?

The geologist, ungracefully but gratefully, is permitted (by himself) to ignore the subtle difficulties of definition. With his almost total dedication to history, he is perhaps the epitome of Western man. And yet, through his conception and grasp of the span of Geologic Time, he may romantically appreciate the vision of endless, and to the short-lived organism, seemingly undifferentiated present. (Such is the paradox of the inconsistency of the self-acclaimed rational animal!)

Geologic Time, as opposed to unadorned Time, is not difficult to define: it is, to use Berry's definition, "The time span during which . . . events took place [that are] commonly termed 'geologic' . . ." Since it is the geologist who is authorized to state which events are "geologic," this definition leaves him all the room he may need to maneuver in.

The geologist's interest in time is not academic. As already suggested, his interpretation of earth history through Uniformitarianism depends upon the existence of great enough lengths of time to accomplish feats of considerable magnitude by means of processes that are, for the most part, seemingly infinitesimally slow. Beyond that, the geologist wishes to establish simultaneity or sequence of ancient events, and to state with some accuracy the quantities of time that have elapsed since and between points in antiquity. That is, in the jargon of the science, he is interested in *relative* time and *absolute* time.

Relative time, the placing of events in chronologic order, depends upon one simple idea and one sophisticated idea. The judgment as to simplicity and sophistication is, of course, made in retrospect, and the formal nature of the original presentation of the simple idea makes one suspect that it was not, at one time, obvious.

The simple idea is simply that in a sequence of flat-lying layered rocks any layer underneath another is the older of the two. The assumption here is that the rocks accumulated by deposition, and since deposition must take place *on* something, the something on which the deposition took place must have been there before the material being deposited. Complications arise when the layered sequence has been folded, faulted, partially eroded or metamorphosed, or when comparison of relative age between unconnected, separate sequences of layered rocks is attempted. Early workers felt that similarity in rock appearance was sufficient to establish equivalency of age, that rocks of similar mineralogy and texture formed at the same time, and that "older-looking" rocks *were* older than "younger-looking" rocks. Later it was realized that similarity in appearance indicated similarity of conditions during formation of the rocks, and that such similarities might occur repeatedly throughout Geologic Time and could not be used to suggest time equivalencies in separated sequences.

This difficulty was overcome by the sophisticated idea. The latter was formulated by an Englishman, William Smith, at the end of the eighteenth century. Smith, who was well acquainted with the layered rocks of Britain through his duties as canal builder and engineer, noted that they often contained abundant organic impressions and petrified remains. Moreover, and more significantly, he noted that each level within the layered rock sequences contained an assemblage of these fossil organisms that was characteristic of that level and distinct from other assemblages above and beneath it, even though the rocks containing them might be similar in other respects. It thus became possible, through examination of fossil assemblages, to correlate (establish time equivalency between) rocks that were geographically widely separated. Even though it was subsequently realized that different fossil assemblages might be found in rocks of equivalent age (for instance, an assemblage of bones of land animals would be quite distinct from an assemblage of sea shells that accumulated at the same time), a powerful tool for deciphering geologic history had been developed. In short, Smith, thought of as a common man in his time (because he worked for a living), had come up with an uncommonly good idea.

As to why the fossil assemblages varied vertically in sequences of layered rocks, that was not yet understood. Indeed, it was still not uncommon to encounter the belief that fossils were placed in the rocks by evil powers, that they were discarded failures in God's attempt to create viable creatures, or that they grew in the rocks through the irradiative powers of heavenly bodies (a fascinating account is given by Adams in *The Birth and Development of the Geological Sciences*).

By the middle of the nineteenth century Charles Lyell, after careful examination of many fossil assemblages, noted that in certain sequences of fossiliferous rock, those towards the bottom contained relatively few organisms similar to those living today, while those nearer the top contained greater percentages of species still extant. There was, evidently, evidence of a progression or evolution through time of life that was increasingly similar to the flora and fauna of today. Explanation of this evolution was provided in Uniformitarian terms by Darwin in 1859, and the vertical changes in fossil assemblages that Smith noticed took on profound implications in religious, philosophic and scientific spheres, as elsewhere suggested.

Through geologic application of the "simple idea" and the "sophisticated idea" (known respectively and more formidably as the Law of Superposition and the Law of Faunal Succession), it seemed possible, by integration of information gathered from many areas, to create a theoretical sequence of layered rocks that would represent material deposited from earliest times up to and including the present. This sequence, called the "Geologic Column," would, in fact, represent (but not be) Geologic Time. The layers towards the bottom would represent the most ancient times, and the ones towards the top the most recent. Subdivisions of Geologic Time might be made on the basis of significant changes in the fossil assemblages incorporated within the Geologic Column. New outcrops of layered rocks might be placed in their chronologically correct place by determining which rocks within the Column were their equivalents—not always an easy task, but ideally theoretically possible.

The Geologic Column as the ultimate chronicle of Geologic Time was undoubtedly a marvelous idea. It seems fitting that the "sands of time" should not be permitted to fritter away, uselessly, when they might be piled up into something sensible and orderly. Several problems remained, however. While it is true that most continental areas are covered by layered rocks, and that many of these are fossiliferous and fairly flat-lying, the great internal mass of the earth's crust, and quite large surface areas, are underlain by unfossiliferous and/or unlayered materials or by layered rocks whose layering is either not due to deposition, or if it is, has been too contorted and altered to be meaningfully equated to the Geologic Column. Indeed, the bulk of Geologic Time is represented by rocks formed before the appearance of abundant and preservable life. In most cases, they have been intensely metamorphosed, melted or intruded by molten material. How can these rocks be related to each other and to the more simple, fossiliferous rocks? And how old are the simple, fossiliferous rocks themselves?

The solution lies in the concept of *absolute time*. If relative time is difficult or impossible to work out according to the stratigraphic principles of superposition and faunal succession, assignment of actual ages, in terms of numbers of years before the present, would establish chronologic order. Similarly, assignment of absolute ages to rocks underneath or within the orderly, fossiliferous part of the Geologic Column, would permit study of rates of evolution, deformation, sedimentation, erosion, etc.

The determination of absolute age has, of course, its own intricacies. What is needed is a process whose progress may be measured accurately, whose rate is unvarying throughout vast lengths of time and under extreme conditions, and whose initiation coincided with the genesis of the material or event under consideration. These seemingly impossible requirements appear to have been met by the process of radioactive decay. Certain chemical elements gradually, unceasingly deteriorate at determinable rates into certain other inert "daughter" elements. When such identifiable daughter elements and their radioactive parents are found in a rock, and the rate of conversion is established, then an estimate may be made as to when the deterioration began. The assumptions made here are that the rate of conversion was constant, and that none of the substances involved have escaped.

But where are such elements found? And did the radioactive decay coincide with the genesis of the rock? And have there been no leaks? The first question may be answered in a comfortably general way by saying that many igneous rocks, a number of metamorphic rocks and a few sedimentary rocks contain the appropriate materials for absolute age determination. As to when the radioactive clock started, each case must be considered separately. In igneous rocks, the hands were set at zero when the melt from which the rock formed solidified; thus the determined age represents time elapsed since crystallization of the parent magma or lava. For sedimentary rocks, the age determined may be that of the parent rocks from which the sediment was derived or of minerals formed during or after the sedimentation process—so you have to be careful to know how the rock was formed and which part you are measuring. When rocks are metamorphosed the problems are compounded. Previously started clocks may "leak" or may continue undisturbed, and new clocks may be set for the first time. Again, you have to "know geologically" the material to which you are applying these technically sophisticated techniques. That is, for any age determination it is important that the investigator be fully aware of the possible historic complexities of the rock under consideration, so that he may know

Natural Tank, Waterpocket Canyon, Capitol Reef National Monument, Utah

what significance and reliability to place on his results. (As always in trying to solve a problem, it is helpful to have an idea beforehand what the solution might be.)

On the surface, it is the absolute age of events that is most stunning. The billions of years since the accumulation of the earth, the hundreds of millions of years that have passed since a given pile of bones walked around inside its envelope of living flesh, these are the figures that produce the gasps of awe. But relative time has its own impressive aura. It is interesting, after all, to trace the sequences of changes in the earth and in living organisms, to see from what and where man has come and to note the high percentage of species that have flourished only to become, seemingly forever, extinct. It is, perhaps, relative time, awareness of chronologic order, that makes Western man ponder pre-Western man, and wonder who or what will follow us.

Collecting and Collectors

Mr. Boyle keeps a spittoon in his rock shop and makes good use of it. The spittoon or a turning politely towards it, punctuates his conversation and his sales. Mr. Boyle's shop faces the southern Rockies. The specimens on his shelves come from there and, beyond prejudice, from the rest of North America. Indeed, any hint of provincialism is dispelled by the labeled presence of material carted in from the mines and fields of Europe, Africa and Asia—though of necessity only symbolic in number and size.

Like most rock shops, Mr. Boyle's keeps not rocks but minerals, together with cutting tools, grinding wheels, a few fossils, some faceted gems, a collection of old glass bottles that have turned purple through long exposure to desert or high mountain sun and one or two lamp bases and tabletops constructed from slabbed specimens in his spare time. The absence of rocks in a rock shop is not an oversight but a matter of misnomer. Amateur mineral collectors call themselves "rock hounds," and "rock" shops are logically where they trade and buy.

As Mr. Boyle told me, rock shops sprang up across America—together with the motel—in abundance after the end of World War II. It was the reappearance of gasoline, the possibility of once again buying automobiles, and cash to do it that put Americans on the road. And, as I can testify and Mr. Boyle affirm, a traveling American is an acquisitive animal. He likes to bring back something, preferably exotic, from where he has been. "Rocks" for many seemed to be satisfactory mementos; hence the birth of a multi-million dollar industry.

My own collection of minerals began before I had ever seen or heard of a "rock" shop. Each specimen I had was collected from the mine, roadcut or quarry where it originated, and carried with it the memory of early rising, hours of travel, the excitement of discovery when it was at last uncovered from under a pile of rubble or chopped from a wall or cliff face. To buy "rocks" would have seemed a violation of some basic ground rule: the prize should be the fruit of labor directly associated with it, the culmination of search, fresh air, and a measure of luck.

It was, therefore, a long time after their appearance that I first went into a rock shop. To my surprise I found them a distinct comfort. The comfort stemmed partly from the material on the shelves and partly from the personages behind the counter. Owners of rock shops seem to be endowed with a geniality whose source I can only speculate upon. Show an interest in the specimens and the gleam in their eye is more often pride than the prospect of a sale. Indeed, the best specimens are usually not for sale. They are there to admire, to handle reverently, to chat and reminisce about.

There is a touch of magic about entering a rock shop. When you cross the threshold you leave where you are and enter a realm that is always familiar, always the same. The siderite from Roxbury, the geode from the Yucatan, the fluorite from Cornwall greet you from the shelves and trigger chains of association. There is an intimacy in their presence, a link with the past, other places and people that is overwhelming and undeniable.

As I suggested previously, to owners of rock shops the distinction between minerals and rocks is academic and almost universally ignored. But it is, nevertheless, real. Minerals and rocks are physically and aesthetically poles apart. Furthermore, the differences between them are reflected in the personalities of their collectors.

Technically, minerals are a simpler class of objects than are rocks; they are, in fact, what rocks are made of. Minerals are chemical compounds; rocks are mixtures of these compounds. In order of increasing complexity, the inorganic world is made of atoms, minerals (natural compounds) and then rocks (natural mixtures of minerals in some manner cemented or grown together).

More formally, a mineral is a naturally occurring, inorganic compound, with definite chemical and physical properties that are the result of the orderly, regular, internal arrangement of the atoms. That is, a mineral is defined not only by the materials that compose it, but also by the three-dimensional pattern that the atoms form when they combine. This pattern, called a *space-lattice*, is arrived at through the accommodation of the combining atoms to each others sizes and electrical charges, and results in structural stability. Indeed, minerals with similar arrangements of atoms but different chemical compositions may be more alike than minerals with similar or identical compositions but different "space-lattices."

Although they are not likely to all appear in any rock shop I know of, roughly six thousand mineral species occur naturally in the crust of the earth. Only about *one hundred* of them, however, might be termed common or abundant. These figures are surprisingly low if you consider that the number of kinds of atoms (elements) alone is over a hundred, and combinations of elements in various proportions could theoretically number in the millions. The answer to this dilemma is that the number of *really* abundant chemical elements is less than a dozen and the number of *fairly* abundant elements, perhaps two dozen. This general unavailability of most elements, together with the limited proportions in which they may combine and meet the requirements for building space-lattices, accounts for the relatively restricted number of minerals.

From the point of view of someone who is interested in a

complete collection, six thousand varieties probably presents a reasonable challenge. Actually, this number becomes larger for the connoisseur. Mineral species often bear the stamp of their occurrence in a manner that is unmistakable. Siderite, for instance, may occur in dozens of localities around the globe, but specimens from the now unused, pre-Revolutionary War iron mine at Roxbury, Connecticut, have a range of colors, lusters and shapes that are somehow unique. The true collector will want to have these, together with siderites from all other important locales in his collection.

The prize specimens of most collections are crystals. Crystals form when a mineral grows in an open, permissive environment that permits the regular, internal structure to express itself as a regular, geometric, external shape. Such environments include air, water, cavities in rocks, molten lava or magma, or in some cases, the actual interior of solid rock. But such "free" environments are rare and preserved and available crystals, even rarer. Most minerals grow in a restricted environment and their shapes are the irregular ones impressed upon them by the walls of the space in which they grow.

The increasing rarity of crystals—the rarity increases as the number of new deposits decreases and the number of collectors increases—suggests that as the years pass the focus of mineral collecting must change. The emphasis might change from crystal collecting to a more general interest in variation within non-crystal specimens. Actually, the latter are often quite as spectacular as their more perfect brothers, but I suspect that as crystals are harder to obtain and their prices increase, the casual tourist trade that many rock shops depend on will decline. In that event, rock shops will not disappear; more likely, the merchandise will diversify to include old bottles, artificial slag, local "art" objects and possibly popcorn. The rock shop will follow in the footsteps of the drug store.

There is another alternative, however, for those who like to collect—to actually be a *rock* collector rather than a mineral collector. That rock collecting is not popular is not mysterious. Rock, by definition, is a coherent, mutually attached agglomeration of minerals. It may be composed mainly of one mineral or may be a mixture of several, usually less than half a dozen. As is often the case, the whole is quite other than the sum of its parts. Unless the minerals themselves happen to be either crystals, highly colored or otherwise spectacular, rock is generally diffuse and undramatic. Its attributes depend more on its overall shape, slight variations in color and texture, contrasts of lights and darks, cavities and protuberances.

Rock appreciation, therefore, is less immediate than mineral appreciation. It is perhaps an acquired taste. If rock *collecting* is not common, rock *arranging* is. In Japan, *bon-seki* or stone arranging is a highly developed and respected art. For the stone arranger, the quality of a rock lies not only in itself but also in its relationship to it surroundings. One of the keys to stone arranging is restraint. A handful of boulders may lie in a sea of sand or pebbles. The Western temperament may pause to admire *bon-seki*, but probably will not find that stone arranging assuages its thirst for acquisition and activity. Rock *collecting* may be the answer.

Most rock collectors at present are geologists. For them the apparent dullness of most specimens is mitigated by the genetic marks they bear. Pyroclastics hurled from volcanoes, ropy lavas, frothy pumice, intricately weathered sandstones are among the more dramatic collectors' items. Prosaic granites, basalts, limestones and schists are more apt to gather dust than compliments from the layman.

What is the difficulty? The problem is one of display. Even the most indifferent observer will agree that a *cliff* of granite, a *mountain* of limestone, or a *ridge* of basalt are of different cloth from the dull, labeled lumps on a shelf. It is the removal of the rocks from their natural environment that is the aesthetic catastrophe.

The solution, perhaps, is to re-create rock environments—in miniature. This is possible as most rocks possess no inherent scale.

(Overleaf) Monument Canyon, Colorado National Monument

Rock separated from objects whose size we know can be any size. Set aside an area—indoors or outdoors, it doesn't matter—and create. Build microcosms—deserts, mountain ranges, jungles, seashores. Invent strange hybrids. Add cactus, narrow-gauge railways, towns, people. The possibilities are endless. If you wish to know them, the names and scientific value of the specimens can be your own secret knowledge or recorded elsewhere. If one justification of collecting is the conveying to others of a certain joy, these minute, rocky worlds will do the trick better than any linear, orderly collection, no matter how well typed the descriptions or sparkly the cabinet glass.

Am I trying to undermine rock shops and their success as a multi-million dollar industry? Not at all. Probably rock shops will never sell many rocks, but I suspect their proprietors are as inventive as they are genial. Furthermore, if Mr. Boyle is a good example, most of these gentlemen came to open rock shops only after many years of climbing the hills and searching the wilderness. They have a love of the substance they deal with; to them the idea of creating rocky worlds would not be alien. It is, after all, only a return to childhood, a closeness to the earth, an attempt to have part of the world as we would wish it.

Where the Ideal Exists

For those who collect minerals (rather than rocks), crystals are inevitably the prize specimens—and it's not hard to understand why. A crystal is such a simple approach to perfection: flat planes meet each other at sharp points and straight edges; light, unwary, bounces around the interior, trying to escape; facets gleam, unabashedly answering the salute of the sun. Once formed and complete, the crystal is an impregnable fortress, moats full, drawbridges raised, no unauthorized entry possible. It floats with us through the ether, reminder of the frozen absolute waiting above the clouds; its symmetry stands as a vague rebuke to our own uneven contours; drawn tight, its icy cloak brooks no queries—until it is cracked by frost or crushed under the hoof of a wandering beast.

But crystals, albeit reluctantly, reveal some of their secrets if examined carefully enough. A close look and their perfection disappears, replaced by flaws and complexity, perhaps ultimately more endearing. Indeed, ideal crystals are Platonic shadows: unattainable, impossible, a pattern to emulate.

Actually, it is deviations from the ideal—at every step in their formation—that permits crystals, as we know them, to exist at all. Crystals form as substances abandon the molecular chaos of liquid or gassy states for the more sober realities of the realm of solids. Existence in a non-solid state is a fast-moving dance. Prodded by falling temperatures, changes of pressure, and a variety of other intangibles, molecular particles must exchange free-wheeling disorder for the sluggish dignity of assigned positions in space—about which they may wiggle and jiggle, but which they cannot leave. They occupy points in a space-lattice which is, as mentioned previously, a three-dimensional regularly repeating pattern that extends outwards in all directions. The external regularity of the crystal that we see is only an incidental and accidental manifestation of the mathematics of the space-lattice.

At the birth of a crystal a discrete number of particles join together to form incredibly small polyhedrons: cubes, pyramids, prisms. It grows, if it is going to grow, by the addition according to formula of rows and layers of particles to each of its faces. Larger and larger it grows, its eventual size curtailed only by the supply of material, the stability of the environment and the walls of its container. It all sounds very simple and straightforward. It is and it isn't.

Consider the difficulty of initiating the space-lattice or *crystal nucleus*. The fast-moving particles in the parent liquid or gas must join together. Such a linking can only take place when the electrostatic attraction that the particles have for one another equals or exceeds the energy of motion they possess. If they are moving too fast, they veer towards each other as they pass, but they will not fasten together. If they are moving more slowly, they may succumb to their mutual attraction, join together and initiate a space-lattice. Space-lattice formation, nucleation, the start of crystallization requires, therefore, relatively low average particle velocity. But even if the *average* velocity meets theoretical requirements, some few particles may have *above average velocity* and be still moving too fast. In fact, if they encounter the infant crystal, instead of joining onto it, they may well smash it to pieces.

The problem is somewhat akin to trying to "triangle" slightly sticky billiard balls by tilting a billiard table back and forth—hoping that in their chance, brief encounters the balls will stick together rather than bounce apart, and hoping, perhaps even more fervently, that whatever part of the "triangle" has already formed will not be broken up by the next stray that crashes into it.

Well, you might say, then it is simply a matter of waiting until the average particle velocity is sufficiently low so that freak, high-speed encounters become statistically unlikely. For some substances such waiting bears fruit. For other substances, however, the effect of lowering the average velocity beyond a certain point is the "freezing" of chaos. That is, the particles more or less stop where they are, too sluggish to form a pattern. They have responded to the grosser immediacies of strong attraction, neglecting the niceties that would tug them gently into "correct" positions. They have become what is known as a "glass." It's as if someone poured glue all over the billiard table.

These difficulties of nucleation are most acute when the parent substances are pure; the presence of "foreign" particles often solves the problem. Rather than attempting to initiate a space-lattice or form a seed crystal themselves, the combining atoms may fasten onto a speck of dust, a crystal of another species, or even an irregular projection on the wall of the container—and grow out from there. They solve the problem by avoiding it. Foreign nucleating substances can often be seen at the center of a crystal—testimony to devious character, devious from birth!

But even if its start was "pure," the growing crystal encounters the same problem again at every stage in its growth. Once a given layer of the crystal is complete, particles approaching it to form the next layer have difficulty in finding a place to attach themselves. Like mountain climbers attacking a sheer rock face, their progress depends upon the presence of imperfections: projections or cracks. Luckily, from the point of view of the growing crystal, such imperfections are common; they result from the incorporation in the developing space-lattice of particles that do not really belong there, actual holes in the space-lattice, or slight shifts in the direc-

tion of the pattern of the lattice. These irregularities—the result of rapid growth—are what ensure the crystal's future.

Space-lattices, with their inherent possibilities of achieving regular external boundaries, grow in a number of distinct environments. Within some of these environments, growth is limited only by the supply of materials: development of crystal form is the rule rather than the exception. Snowflakes, for example, grow surrounded only by air and cold. An infinity of perfections is the result. Other crystals develop within empty cavities in rock. Unless they are overambitious and fill the entire space—in which case they will mimic the shape of the cavity—they too will be bounded on all sides by perfect faces, except where they are attached to a wall or each other. These "open" environments, however, are exceptional. Most space-lattices develop either in a cooling or evaporating liquid, or else within the actual body of solid rock.

In the first instance, the earliest space-lattices to form encounter only liquid as they grow; if they stop growing before they bump into and interfere with each other, they may well achieve and keep their crystal shape. Materials that start or continue to grow in the liquid at a later time must fit themselves into the spaces left between the earlier formed crystals—and usually end up by being irregular.

Many other space-lattices nucleate and develop within solid rock. That is, they form where other space-lattices already exist. Their birth and growth involve the decay and death of the prior crystals: they must successfully dismantle and eliminate the older, pre-existing materials. How and why this happens is somewhat of a mystery, but somehow, whatever material is needed for the development of the new space-lattice obligingly migrates to the growth site, particle by particle, ready to be used. In an equally obliging manner, unwanted atoms of the old space-lattice migrate away from the site, leaving room for the new structure that is replacing it.

Surprisingly enough, space-lattices that grow in this way, under what would seem impossibly restricted conditions, often become

Canyon De Chelly National Monument, Arizona

quite large and even, sometimes, achieve perfect crystal shape. They do have the shortcoming, however, of characteristically being full of inclusions–small islands of old space-lattices that they used for nuclei or that were not completely eliminated.

The ideal crystal, therefore, does not exist. It could not exist. Its very perfection negates its possibility. Its actual manifestation has no more chance than life in a wooden man. What is it then that we admire when we see a crystal? Do we admire the imperfections that made it possible? Or the bulk of the crystal that approximates perfection? It is the latter, for we are impressed by the result rather than the process. In a sense, what we are admiring is specious and doesn't exist except as an illusion that illustrates the grossness of our perception. But this perception *is* our perception and our aesthetic judgments must be based upon it. The only valid question is how to enhance it, make full use of it. Does understanding the process that has led to the present instant accomplish that? In as much as it points towards miracles beyond the senses, it does. Whether we have a sense of beauty is probably independent of whether we have understanding–in the scientific sense. Perhaps without the sense of beauty there is no understanding, for "scientific" analysis without aesthetic satisfaction is barren.

But where does beauty lie? We create abstractions, such as ideal crystals, that cannot exist. Nature creates real crystals that do exist, concretely, as physical entities that can be perceived by our senses. Surely it is nature that approaches whatever perfection is. Real "imperfect" crystals are ideal in that they are possible; our abstract "perfect" ones are really imperfect, impossible approximations. Isn't it, therefore, in the actual world, actual in the ordinary, everyday sense, that the ideal exists–if only we can see it?

Bones and Purple Glass

Purple glass lies scattered about the desert floor. Silicified wood pokes out through dry stream beds. Bleached bones are the clearest evidence of life—except for occasional cholla or tamarisk that mark the empty canyons. When the wind blows, dead cactus limbs rustle and cascades of dust swirl into the air. Otherwise all is still. The sun moves a great angle through the stars between rains. When the rains do come, brief waterfalls are everywhere, boulders tumble down slopes, carried along in roaring sheets of water. New canyons appear at the urging of rushing streams; old ones are deepened and lengthened, their channels a spreading, fingerlike dissection of the earth. In the lowlands, in basins surrounded by mountains, vast lakes shimmer and are gone, leaving behind, when the sun reappears, a blanket of salt and mud cracking in the heat. Flowers may bloom for a few hours. Then, life and movement are dormant until the next passage of rain.

The ecology of life in the desert is transparent. Each change in slope, each level of elevation has its own characteristic population. The shade cast by every pebble or cliff is significant. Life is not haphazard. It chooses very carefully.

In the same sense, the topography of the desert is not accidental.

The occurrence and genesis of every valley, every hill, is revealed in the rocks that lie beneath them. There is little to hide any secrets.

Desert is not an aberration. Indeed, the well-watered parts of the world that more easily meet the requirements of abundant life seem, in a sense, more unlikely—for isn't life itself, in this harsh universe, unlikely?

Deserts occupy great portions of the globe, wherever water is unavailable. They may lie near the sea in the rain-shadows of high mountains, or in the interiors of vast continents where moisture never penetrates, where the breezes have long since lost all breath of rain. Alternatively, absence of water may be due to the pattern of atmospheric circulation. Over wide belts of latitude air constantly descends, receiving rather than releasing moisture.

But in all this the desert is deceptive. Under the driest wastes water is often ubiquitous. Great cities could rise in most desert basins, sucking up the moisture inherited from the past. Ten thousand years ago, during the last Ice Age, the vast Mojave was dotted with lakes. The mountains between the basins were lashed by rain as wet winds flowing south off the glaciers rose and released their bounty. This water now lies stored in the porous, permeable sediments under the desert floors. They are not inexhaustible, but they are considerable. In my lifetime they will remain largely unknown, untapped. I am glad, for I should hate to lose the emptiness of the desert.

The desert floor has a special quality. Most of its major features have been sculptured by water—the absence of vegetation makes every brief, chance storm a catastrophe—but the wind is responsible for many of the minor quirks that form much of its character. The peculiar talent of the wind lies in its ability to separate materials according to size. Water will do this too, but its action is not as delicate. Whole expanses of the desert floor are entirely free of sand and dust; the wind has swept it clean, leaving a garden of pebbles lying on bedrock. Other areas are covered by this same sand, released by the wind when it was slowed by encoun-

Near The Slot, Point Lobos, California

ter with some obstacle. In some places, small oriented tongues of sand lie to the lee of each pebble, pointing out the last direction of a careless breeze. In others, accumulations have built up to great dunes that wander slowly like massive turtles. If water and wind have cooperated, the most barren of all desert floors may be created: endless sheets of bare rock lying open to the sun, washed free of all debris.

Whatever life the desert does support, it includes a breed of people who would live nowhere else. There they find and want the time and quiet to consider—the quality of existence, the nature of change, the presence or absence of God.

The Sinai desert sits between the fingers of the Red Sea, at the junction of Africa and Asia. In its heart is a range of granitic mountains that rise well over a mile above sea level; at their center, highest of all, rise the peaks where Moses received the Ten Commandments. The mountains are part of the Nubian Shield—brown, barren, ringed to the west and north by white dunes that drift over empty plains. Rainfall, when it occurs at all, is in the winter months. Sudden sharp thunderstorms in the mountains fill the unsuspecting wadis (ravines) with water that carries everything before it, a roar that ends only as a patch of damp in the coastal sands, the brief torrents most often dead before they reach the sea. At the base of the granite slopes, however, where they disappear beneath the sediments of the valleys, there is sometimes a blade of grass, a small flower that marks where the rains slid off the mountains into the sand. All the water is not lost. Some has escaped the evaporating eye of the sun, some has not been caught up in the mad rush for the sea. It has sunk slowly through the wadi floor, available to those who search, to those who dig.

The Oasis of Feiran lies halfway between Jebel Moussa (Moses Mountain) and the Gulf of Suez. There, in the desert, seven thousand date palms, manna-bearing tamarisk and spiky acacia grow, clustered along a five-kilometer length of clear stream, carefully guarded and tended by appreciative bedouin. The oasis is isolated from the world, reached only by forty-seven kilometers of uncertain road that parallels and crisscrosses the Wadi Feiran, a dry stream bed that runs inland from the coast. Past the oasis the wadi bed is dry again and rising eventually joins Wadi El Sheikh. For fifty kilometers more the road twists with the empty stream, passing through narrow gorges, meandering crookedly across flat, dusty intermontane basins. Coastal sediments have given way to dark gneisses and schists; basalt-spined ridges appear and are replaced in turn by the high pink masses of the central Sinai. Where sheer granite cliffs rise thousands of feet on all sides, the road gives out at last, ending at the green monastery gardens of the Oasis of Saint Catherine, huddled at the base of Jebel Moussa.

The buildings of the monastery date in part from the sixth century. They have remained well-preserved in the desert solitude. At one time four hundred monks resided there; now it lies almost deserted, empty except for perhaps a dozen Fathers and their servants. Throughout the ages the waters in the oasis have risen without fail. The monks, however, look upward, not down–up, to where, amongst the clouds, Moses spoke with God. A stairway has been constructed, four thousand three hundred granite slabs that wind to the top of the mountain. Halfway, the stairs pass through the gateway of Saint Stephan, who until he died in 580 A.D., questioned pilgrims as to their motives and established their purity. At the top itself, open to the sky and the wind, are a small chapel and mosque. On all sides, barren mountains, empty valleys, dry wadis fall away. Almost invisible ruins of other ancient monuments can be seen on the facing peaks–tiny fly specks hanging in the immense spaces of the desert.

At the bottom of the mountain, at the edge of the monastery, is a small building that contains the bones of all the monks who over the centuries have lived and died there. Three thousand skulls rest stacked in one corner of the ossuary; three thousand headless skeletons lie in a bin next to them. Saint Stephan, mummified and black robed, sits in a glass case guarding them, an eternal smile on his face.

Tradition has it that at the end of the world the heads will be reunited with their bodies. This may be so, but the distance they will have to travel will probably be more than just the width of the ossuary. How long will it be until the building crumbles and this charnel heap is spread over the valley, bone after bone making its own slow journey down the wadis, back to the sea?

Why do men—not born there—come to live in the desert? Is it the immediacy of things, the absence of intervening subtleties? In the Sinai religious motivation offers itself as a clear answer. But why was the Law given to man in the desert. Is it an isolation, lack of distraction, a silence that permits an inner self to be heard?

Across from the Sinai in the desert of the Negev, at Ein Boqeq on the slopes of the Wadi Arava, a group of Americans are trying to establish a colony. Drive the road from Be'er Sheva to the south and you can meet them. In an effort to raise capital they run a coffee shop—selling pancakes, home-made pie, tea and coffee to the truck convoys of the military. Eventually, they hope to buy cattle and grow grain. Officials from the government come and tell them what they want to do is impossible. Everyone who passes agrees. But they watch the water flow from their small well and gaze out over the scorched alluvium towards the mountains of Edom. The land is empty; they want to make it theirs.

At Silver Peak, Nevada, the desert is equally dry. The only sure livelihood is the removal of salt from the adjacent desiccated lake bed. Yet people quite uninvolved in this enterprise live there too—in trailers, in mud huts, without running water and, until recently, without electricity. They are not quite destitute; they could move if they wanted. But they don't. Instead, systematically they stake out claims in the nearby hills and with backbreaking effort put in the improvements legally necessary to keep the land. There are easier ways to make a living. But one man said, and his wife agreed, "We wouldn't live anywhere else." The mountains shimmer with heat in the distance. By nine in the morning it is 115 degrees in the shade.

Perhaps it is an instinct to impose human patterns upon what seems chaos that brings these people to the desert. But perhaps it is the absence of humanity that permits for those who live there the discovery that in nature lies the least chaos.

Whatever the reason, wherever the desert water seeps to the surface, man puts up his tents, leaving, eventually, his bones bleaching in the sun and glass bottles to turn purple beside them.

Things as They Aren't

When we see forms in rock—and who doesn't?—the forms spring from the mind as much as from the substance. But how can rocky form be described? How can it be experienced? What words can be used that are in any sense relevant to the stone involved? What geometry encompasses the quality needed to define the essence of granite, of limestone, shale or schist? We are accustomed to think in terms of slabs, of spires, fins, blocks, walls, sheets and arches, of basins, bowls, fluted gullies, bridges and columns, but it is our minds as much as erosion that creates these. Rocks cannot speak, at least not in simple, verbal symbols. They can point to process or to the past or to the imminently or remotely possible. They can hint at change or demonstrate it catastrophically. But the momentarily actual, the concrete manifestation that sits—solidly or delicately—astride the boundary between past and future, what can that do or say, except be?

So we are left with a problem—a problem because involvement with rock necessitates, on our part, reaction. There is the rock. It has a shape, a size, a presence. We cannot ignore it. We don't want to ignore it. In some fashion we must deal with it. Good. We can climb it, perhaps pick it up and put it in our pocket. With courage we may make it the object of our canvas or our camera; with peace we may lie in its curves or tuck ourselves into its recesses and niches. But at some point—for that is our nature—we are impelled to use words.

What words? Do we describe a rock formation as the Small-Boulder-on-a-Large-Boulder? As the Three-Pillars-Red-on-Top-White-on-Bottom-Joined-Midway? Certainly. But descriptions are arbitrary, lengthy, and soon give way to names. Wherever forms of the earth are found that are discrete, spectacular entities, unusual or majestic in some fashion, they soon receive the additional accolade of designation. Sometimes the names are simple, such as North Point, Grand Rock, Henry's Gulch, McCullough's Downfall or the Squeeze, recording location, description, ownership, event or physical necessity. Such names are passive, inoffensive and make no attempt to claim the glory of what they represent. But other names are usurpers: they obscure rather than indicate, shape the earth to fit some Procrustean image rather than illuminate and reveal intrinsic form. A rock called North Point or McCullough's Downfall, unless you know McCullough, remains itself and open to each new viewer, but a rock dubbed the Whale or Cinderella can never again be quite what it is, free from fishy line or girlish flounce. Follow, for instance, some well-meaning, usually much-appreciated guide through the intricate passages of an underground grotto: "On your left, ladies and gentlemen, see the Three Nuns stepping out from the Cathedral. And over here—watch your step—an old friend, Donald Duck, sitting around the Campfire with Snow White and the Seven Dwarfs. Follow me now through the Devil's Garden into the Belly of the Whale. See, there's Jonah Holding a Lantern, looking for the way out . . ." a torrent of chatter and cheap fantasy to describe stalactites and stalagmites, dripstone formations that accumulated silently, without human aid or attention, over thousands of years. What a shame such constant analogy is considered entertaining or necessary. It is true that analogy makes the unfamiliar familiar, relates it to things we feel at ease with and understand. But why do we need such protection and isolation from our surroundings when no attack is expected? Surely within the folds of a well-explored cave no artificial heightening of drama, no false sense of danger is required to attract our attention or to keep us from being bored. Given half a chance, the quiet, glistening world of the cave will weave its own incomparable spell. The guide's patter can only be distracting—but perhaps it is meant to be, for we have lost our sense of the religious, the numinous. Somehow we never let ourselves get beyond being uncomfortable when faced with the mysterious or powerful—we giggle nervously and try to reduce it to the mundane.

But finding forms in rocks is as old as humanity. For the ancient Greeks, the hills were peopled with gods; indeed, the hills were gods. The Breasts of Venus emerged as islands in the Gulf of Argolikos, the summit of Mount Ida was upheld by the Bull of Minos. But there is a difference between Donald Duck and the "Breasts of Venus": the ancients believed, whereas we don't; we are just being cute—or uneasy. A mountain concealing the Bull of Minos or created as a result of the anger of Zeus really was such—with practical and moral implications. For the ancients the earth was something lived with in intimacy and ignored at their peril; it was not something to be dealt with lightly. In our age of "enlightenment," however, we erect a shield against the wonder and impact of the earth and call it "sophistication," believing, in our ignorance, that such ignorance is knowledge.

Sometimes man goes beyond the creation of form in rock through analogy to do so in actuality. At Mount Rushmore a sculptor converted the cliffs into giant likenesses of the heads of Presidents. For him there was undoubtedly contact with the earth—an embrace, savage but legitimate, through hammer, drill and dynamite. But for the thousands who visit the site of his work, set aside as a national monument in the Black Hills of South Dakota, there can be only passive admiration of the accomplishment, the product, rather than active participation in what must have been his sense of the mountain and the rock. Indeed, his

Canyon De Chelly National Monument, Arizona

monumental sculpture is often considered a victory over the earth. Alas, or happily, this victory must eventually decay and crumble back to the natural silhouette of the hill out of which it was carved. Perhaps it would have been better if the last act of the sculptor had been to destroy his work and let the mosses and lichens cover the scene of his struggle. Such an act would have inspired more profound thought and reflection than does contemplation of rocky, corpselike images of former American leaders.

Why can't we accept things as they are? Why can't the mountain remain a mountain, and the guide say nothing—just guide, bring us to the place and give us time?

In the days of Queen Victoria, the legs of chairs were dressed in ruffles so that they should not be naked. The highest form of expression was euphemism. Have we, today, progressed very far? The ruffles are gone; the curlicues are gone; we take pride in "telling it like it is." But do we? Our buildings rise in sheer planes, our streets follow straight lines, unrelieved by mystery or surprise. Is that how things are? Then why is it that people mutter in despair? Why is there the urge to curl up in the dark earth or retreat to the bottoms of secret canyons? Are hidden passages and strange things around corners really anachronisms? If it is true that in architecture, in sculpture, in environmental design, materials should express themselves as themselves, then a closer look at the surrounding world is badly needed. Being "modern" we can agree that rock as rock is not flesh or floral bower, but then neither is it rectangular slab after rectangular slab; it is somewhere in between, an organic self. Unmasked, unsimplified, it has weight, form and texture, which is for each person a unique experience: one of touch, of heft, of even taste and smell, but always rooted in the specific realities of earth and stone. The variety of rock is infinite but circumscribed by process and substance. It may suggest eternity, but it is constantly being created and constantly being destroyed. It is, at each instant, the summary of its past and the threshold of its future. What we sense as stone, therefore—its solidity, its massiveness—is an illusive flicker in a blur of change—division, cohesion, concentration, dispersion. Each rock is a moment of time, a pocket of space—and is a sharp comment on our own fragile accident of life. But as it is our opposite, so it is our compliment. Through contrast it gives us dimension; it balances the scale. What more can we ask of it? If we impress our own images upon it, by that much we lose it. If we call it or make it something else, we destroy that portion of it that exists within us. But with us or without us, rock itself will remain, and that is an ultimate comfort.

Parametric Roulette

There are certain variables whose values control the world and which, if they were different, would result in a universe unlike anything we know. Consider, for example, the pull of gravity: If the strength of that pull were substantially changed, then we and everything around us would be correspondingly altered. Indeed, if speculation were to attempt logic, changes in the pull of gravity would result from changes in the earth's mass, which in turn might suggest changes in its composition, structure and so on. In brief, it might very easily become a planet on which life could not exist. But suppose the pull of gravity were to change without any change in the mass of the earth or of the object upon which such pull was being exerted? Instead, suppose that it was the value of the gravitational constant itself that were to change; that is, that gravity on the earth *as we know it* were to somehow increase or decrease.

Variables such as the strength of a gravitational field are often called by the more sober term *parameters.* Parameters cannot only be made larger or smaller, but can change from positive to negative. (In the case of gravity, attraction would become repulsion.)

Changes in the values and signs of parameters are, of course, not necessarily possible, but they are an interesting way of looking at things. When you consider that slight alterations can have most drastic effects, it seems as if the world as it is, is almost an accident.

For instance, most substances shrink when they change from liquid to solid. But water, (almost) as if by some "accident," is an exception: water, when it changes to ice, expands, and as a result, ice floats in the water from which it was formed. Imagine, though, what would happen, if, by some different accident, water was not an exception to the general rule. The consequences would be numerous: ships, for example, would have to avoid submerged ice reefs rather than more or less visible bergs; fish could not exist in lakes or inland seas in cold climates, for water-filled basins, upon freezing, would become solid from the bottom up. (In the world as it is, ice, when it forms on the surface of a lake, insulates the deeper, warmer waters; but if ice were dense and sank after it formed, it would continually force the warm water to the surface rather than protecting it, and lakes would rapidly become completely frozen—as would any fish they contained.)

In a way, however, this game of "accidents" is silly, for it assumes that one relationship can be changed independently of everything else. It is quite possible that if ice *were* denser than water, evolution might have produced warm-blooded fish able to hibernate in a frozen state, or keep themselves surrounded by pockets of unfrozen water.

But the game is fun to play—in this arbitrary manner—because it reveals the fragile sequences of events and accidents of relationship that have resulted in the world as we know it. If, to continue, water did not expand upon freezing, the whole character of landscape would be subtly different. In cold regions, one of the major factors in the breakdown and shattering of rock is the freezing and subsequent expansion of water caught in cracks. Frost wedging, as this process is called, rapidly reduces seemingly impervious stony masses to angular rubble which then is capable of being

easily removed by gravity, ice, wind, water. If frost wedging were to cease, both weathering and erosion in cold areas would be drastically reduced. Temperate and polar land regions might be characterized by unusually high average elevation but low relief. Little sediment production would occur; streams would run clear and have little power to erode. Creep, the slow ubiquitous downslope movement of soil and loose materials, would be considerably reduced in the absence of the effects of freeze and thaw. Angular talus piles, boulder strewn fields, rocky brooks, so much of the paraphernalia that springs to mind when we think of wintery places would be minimal or non-existent.

In warmer and tropical regions, rock disintegration is the result of frost wedging accompanied or replaced by the activities of living organisms and the chemical interactions of rock with air and water, and landscape would be much as we know it. Very high mountains, however—those that extend upwards to cold altitudes where life is inhibited—would have a greater tendency to persist, and overall relief in warm regions might be somewhat greater than at present. All in all, it would not be an unworkable world, but it would somehow not be ours.

Consider another "accident" with more major consequences. The sun is the prime source of heat that controls the temperature of the atmosphere, but peculiarly enough, the troposphere or lower atmosphere, which contains the bulk of all air and in which we live, depends upon the surface of the earth for its caloric intake. That is to say, radiation from the sun passes through the troposphere essentially without heating it. Instead, it heats the surface of the earth, which, in turn, heats the lower atmosphere. The result is significant: the lower parts of the troposphere are warmer than the upper parts. Everyone who has climbed from a low plain to a mountain top is familiar with this temperature change but not, perhaps, with its importance. In fact, it is this temperature distribution that accounts for much of atmospheric circulation, weather, erosion and mountains and plains as we know

Arches National Monument, Utah

them. Very simply, warm low-density air is overlain by cold high-density air—and they tend to change places. In short, the lower atmosphere is unstable and involved in constant convective overturn. If the troposphere were heated directly by the sun and was warm at the top and cold at the bottom, it would be stable and motionless, except for the frictional effects of the earth's rotation and tidal pulls of the sun and moon. Patterns of precipitation would be radically different or non-existent. Rain and snow are the result of the cooling of water vapor as it evaporates and rises. If, as it rose, it became warmer rather than colder, it would have no tendency to condense or freeze, except perhaps when it passed out of the troposphere into the higher reaches of the atmosphere. Indeed, water might not exist on the face of the earth at all, for once it evaporated it might never return. There would be, instead, a massive high-altitude cloud layer above the troposphere, and day and night would be evidenced only by alternations between dusk and dark. Anyone visiting such a gloomy planet might hardly be interested in noting that there would be no rivers, glaciers or oceans to erode and transport rock. Sandstones, shales and limestone would not exist. Metamorphic rocks or intrusive igneous rocks might exist but would be rarely seen at the surface of the earth for there would be little mechanism capable of stripping away overlying material. Only lavas and ash would be visible. Volcanic eruptions might provide brief intervals of flaring brightness; wisps of fumarolic steam might rise from crevices on occasion; but except for material slumping from shifting fault scarps or folding strata, all would be quite lifeless and still. Erosion would be restricted to that caused by gravity and whatever winds stirred the cold heavy air. Anticlines, synclines and other structures imposed upon the volcanic layers would long remain unbreached and perfect. The world would be a place of rhythmic, geometric forms transfixed in shadowless half light. Clouds would be eternal, but the only rain would be the imperceptible influx of meteoric and volcanic dust and the chance crash of falling stars. Real stars, the sun, the moon, all would be unknown.

If the "accident" of density inversion were eliminated on a more general basis, the bleak world just described would become even more so. A temperature distribution and consequent imbalance similar to that in the troposphere is most likely present in the upper portions of the earth's mantle. The earth's mantle forms its major bulk and consists of white-hot, plastic, but essentially solid rock that occupies the volume between the crust and the core. Its rate of convective overturn, if indeed it occurs, is probably at the rate of about a centimeter a year, perhaps a billionth of the rate at which air moves. If this overturn, too, were to cease, the motive forces that ensure continued vulcanism, upheaval of mountains, cracking and compression of the crust would probably follow suit. The earth then, except for its astronomic motions, and perhaps a periodic twitching of its magnetic field, would be truly dead.

Not all accidents are quite so devastating. Some just rearrange things. If the earth were at a period of maximum glaciation, with great quantities of what now is ocean water locked up as glacial ice, sea level would be lower by several hundred feet, coastal cities would find themselves scores of miles inland, and climatic belts would be shifted radically. If, on the other hand, the great ice caps were to melt, sea level would rise perhaps three hundred feet, coastal cities would be flooded, and again great shifts in climatic patterns would occur. These possibilities are not really "accidents" in the sense we have been using, for they have in the recent past been reality (although there were no cities to be flooded or stranded). Moreover, glacial fluctuation is most likely now in progress and will continue into the future.

But imagine the political and economic complications if the process were speeded up. In alternate generations Siberia and Alaska might be joined and then separated, Scandinavia buried under ice and then uncovered, the Sahara and Mojave fertile and then desert. Cities on the peripheries of continents might have to be built on great pontoons, so they could follow the constant migration of coast lines.

(Overleaf) Lincoln Gulch, along the Ruby Road, Colorado

For a last turn at parametric roulette, consider what the earth would be like if rocks themselves behaved like ice. To do this would just be a matter of reducing their fundamental strength sufficiently. Ice can support its own weight up to a thickness of one or two hundred feet; accumulation of more snow results in the lower ice spreading out like molasses until the overall thickness again doesn't exceed what its strength can support. Rocks can support perhaps ten thousand feet of their own substance before tending to level out. If rock were as weak as ice, and ice correspondingly weaker, the surface of the earth would be reduced to a flat, almost featureless plain. Volcanic cones would form and then disappear like blisters. Ice, after snow storms, would flow as viscous rivers, temporarily displacing streams. Slight changes in sea level would flood or reveal great stretches of continent.

If a world of weak rocks seems rather dull, if not sticky, then a stronger world might be a welcome contrast. With the mechanical strength of rocks increased several fold, landscape would consist of towering peaks lashed by prodigious rain, snow and howling winds. Huge ice masses would hug the tops and sides of high mountain ranges. This, in short, would be a dramatic world.

In a way, the game of "accidents" is really a formula for instant science fiction. But the real fiction is that parameters could be altered singly in the manner described. If any one of these parameters really changed, they would be just superficial manifestations of much more drastic fundamental alterations. That water is more dense than ice under conditions prevalent at the surface of the earth is not an accident but a result of the very nature of atoms and the forces that bond them together. If the water-ice relationship were different, then most everything else in the universe would be different too. To play "accident" at the level of atoms, though, would not be too much fun, for such changes would be too catastrophic to the world we are familiar with. The most strange, after all, is the distortion of the familiar, not its total destruction and replacement. Besides, it would probably require much knowledge and resourcefulness that we don't possess.

Another game related to "accident" might be entitled "purpose." "Purpose" in a way stems out of "accident." In its simpler forms it runs as follows: The world is the way it is because certain "accidents" happened rather than others. Thus the question: Why is the world the way it is? This is a gap that many are anxious to bridge. Some find in it an opportunity to express their faith in the importance of man: the sun is to provide light by day, the moon light by night; fresh water is so that we can drink; clay, so that we can build. The universe is adapted to man rather than the reverse. To others, man is the accident, temporary no doubt, that has occurred in an obscure corner of an indifferent world.

Perhaps, however, the game of "purpose" is just another way of playing "accident"–a varying of spiritual rather than material parameters. As such, it too is both instructive and arbitrary, revealing and meaningless. But we are, after all, here, so we will–and may as well–play the games.

Dreams of the Earth

One unfortunate aspect of game playing is that there are always some players who insist (sometimes violently) on the unique validity of their rules; those who play religious or scientific games have been particularly guilty. The best antidote, perhaps, to such anti-social certainty, is to consider the benefits of alternate games.

There are in Japan, according to Eugen Herrigel in *Zen in the Art of Archery*, archers who, blindfolded and unoriented, can repeatedly and unerringly score bull's-eye after bull's-eye with their bows and arrows. The training these archers undergo achieves a state of mind where there is no distinction between archer and target, where both are integral parts of some larger framework. In such a situation, the archer has no need to *see* the target; he knows where it is, and the arrow can find it, just as his left hand finds his right.

There are those who claim similar skills with respect to the earth. The figure of the *diviner* or *dowser* searching, forked stick in hand, for water, oil or valuable deposits of ore, is familiar in literature if not in immediate experience. Within geologic circles, the bearer of the forked stick is viewed with a jaundiced eye; most, it is felt, are demonstrably fraudulent, either intentionally or

The Narrows, Capitol Gorge, Capitol Reef National Monument, Utah

unintentionally, and the only professional responsibility of the geologist is to warn the public against them. This is a reasonable attitude—indeed, a reasonable relationship: the geologists shows scientifically that dowsers cannot successfully predict the presence of water or whatever—except where it is more difficult *not* to find it. The public stands warned, but listening to rumors, savoring the occult, it seizes the opportunity of perhaps proving science (and geologists) wrong and employs dowsers anyway. Thus, everyone benefits: the geologist has the chance to feel self-righteous, the public engages in subtle rebellion against a scientific modern world it doesn't really understand, and the bearers of forked sticks earn a living.

Whether there are any unfraudulent dowsers is another matter. If there were a man possessed of special knowledge of the earth, special communication—articulate or instinctual—how would we know him? It might well be that the presence of unsympathetic observers would nullify his abilities. You cannot, after all, force an artist to be creative or a scientist to discover great principles upon demand. The existence of the true dowser, therefore, must remain an open question—an exciting possibility but not, perhaps, something to bank on.

But there are other people who talk to—or listen to—the earth, people who search for more spectacular rewards than water or gold: a glimpse at the future, an understanding of the present, an interpretation of the past. In their ranks must be included geologists, farmers and peasants, those whose eyes and backs are continually bent towards the ground, who learn to read the moods and secrets of the land. But there exist today other individuals who claim, like ancient mystics, literally to hear the earth, who listen for its verbal message and arrange their lives accordingly.

One such man, Ted Lasar, after a flash of revelation, published a short, ambitious, but sincere pamphlet entitled *Life Was Not Always a Tragedy*. The pamphlet stemmed from a moment of insight in which Mr. Lasar realized that the planet Earth is not just the speck in the universe where we live but is itself alive—a sentient, intelligent being with a will and a mind of its own. Furthermore, he discovered that it was this "earth-being," not God or natural accident, that created life and man in particular. Such creation was not whim or passing fancy, but a calculated act designed to accomplish a specific goal. What Lasar learned was that within the not too distant past, comets have repeatedly collided with the earth, each time wreaking terrible havoc upon the planet and its occupants. This premise, he acknowledges, derives from ideas stated in Immanuel Velikovsky's books. He departs from Velikovsky, however, when he goes on to say that these earth-comet collisions not only caused disaster to the people living at the time, but that they also actually caused the earth itself physical pain. It was this pain, he concludes, that inspired the earth to create man so that she might teach him to steer her (earth) out of the path of on-coming collisions.

The earth-being arranged, therefore, that man's early existence be Utopian and geared towards preparing him for his ultimate task. Tragically, however, before man was ready to receive and implement the knowledge of how to steer the earth, comet collisions destroyed his budding civilization. Unable to understand what was happening, man interpreted the awful destruction as divine punishment for his evil ways, and the race was left with deep, permanent, psychological scars. In consequence, all conscious memory of the holocaust was suppressed, as was also all knowledge of the original and primary purpose of man's creation. From then on all men were born and lived with deep-rooted, seemingly baseless guilt and fear that thereafter warped both mankind's personality and institutions. Thus, life was not always a tragedy—but it is now.

However, continues Mr. Lasar, man's salvation is possible—and essential. First he must remember what was the cause of all his ills—earth-comet collision—and understand that the catastrophe was not his fault; that it was a cosmic accident rather than punishment for wrong-doing. Thus, freed from an overwhelming burden of guilt, man could then return to his major task; the steering of the planet. How is this to be done? The earth itself will dictate if man will but learn to hear. Hearing and understanding will come through release of *land.* Land must be bought and then declared unowned; that is, returned to the earth. As this freeing of the land is accomplished, the answer will be given. Man will then have purpose, security and happiness, and the earth will be relieved of pain and the anticipation of pain: a true symbiotic relationship. The alternative, says Mr. Lasar, is repeated earth-comet collision and, sooner or later, the end of us all.

In what manner can such a bizarre view of history be taken? Is it sheer madness or is it truth? A choice seems forced upon us: either his ideas or the main substance of geology, astronomy, archeology, physics and, I suspect, psychology must be rejected. What about Velikovsky's widely read books from which Lasar's ideas stem? They are, in what may legitimately be called responsible professional circles, considered not even heresy, but nonsense. Are Lasar's ideas, therefore, even more so nonsense? Perhaps, except that they contain one important element missing in the Velikovsky genre of pseudo science: a deep concern for the earth and for humanity. If Lasar's thesis were treated as allegory (which he would strongly deny it is), then it would fit right in with what many other people are saying in more acceptable ways: that man must seriously consider his relationship with and responsibility towards his environment. Indeed, it echoes in almost mythic symbols the current sense of crisis and growing public concern over problems of war, conservation and overpopulation.

More prosaically but more effectively, Harrison Brown in *The Challenge of Man's Future*, points out that the success of man's technological civilization has been rooted in the easy availability of raw materials. If civilization should collapse after all surface or near-surface deposits of coal, oil and metals have been depleted (and they will be depleted *forever*, in human if not geological terms, within a relatively few hundreds of years), it will not be able to be rebuilt except, perhaps, on an agrarian, small-population basis.

Lincoln Gulch, along the Ruby Road, Colorado

Barry Commoner also, in *Science and Survival*, has summed up nicely the grim possibilities—or should it be probabilities—of an imminently polluted and overcrowded world.

So how has Lasar fallen outside of this mainstream? Only in the superficial sense of rejecting scientific principles. In considering not only man's physical survival, but also man's human condition—the meaning and purpose of existence, the relationship between what man is, psychologically, ethically, and what his past has been and what his future will be—he goes further than most others. Such a holistic approach is too often considered unsophisticated, simplistic or impossible. Who, after all, can pronounce judgment in so many diverse areas? No one, perhaps, but we all have our hidden thoughts. Lasar has come out with his, and admittedly, they emerge as the ravings of a lunatic or, more kindly, the exhortations of a prophet. Indeed, Lasar's interpretation of reality and his manner of presentation is undeniably biblical: he has direct, revealed knowledge; he has a world view; and he knows right from wrong. In another age he might have found his niche. He has a wonderful all-encompassing grasp of things, even if it is all fantasy. He is a fundamentalist, a preacher of pseudoscientific religion. And what a doctrine—to literally converse with the earth, to consider that our planet needs us as much as we need it, to declare the earth unowned so that it may be heard, to consider man ultimately innocent and capable of finding himself through the earth. What higher form of love could there be out of which man's and the earth's salvation could spring?

Perhaps not higher, but equally as high, and probably more viable in our current age is love of the ordinary rather than the extraordinary. Like most religions based on revealed truth Lasar's lapses quickly into dogma. Lasar, for instance, considers that the chasm in which the Colorado River runs—the Grand Canyon—is the result of comet impact, as is the separation between mesas, the rift valleys of Africa, the rising of mountains, the movement of oceans. That all these features could be the result of common processes operating over great spans of time denies his special knowledge and is, therefore, wrong and an immoral masking of the issue.

But isn't this attitude really alienation from rather than communication with the earth? It stems from abstract revelation, not experience or experiment. Consider the excitement Hutton must have felt as he discovered the simple processes that form the face of the earth, or the elation of William Smith as he pieced together the jigsaw of the rock of the English countryside and assembled the first geologic map. Strong preconceived notions would have been an unsurmountable barrier; their efforts would have been doomed to inconclusiveness. Revelation has its place—the flash that ties previously unconnected fragments together—but if it is taken as a final statement, then it ceases to be enlightenment and becomes dogma.

If anything, the voice of the earth is gentle—not that it cannot be overwhelming at times—and requires a sensitive, open ear to hear it. Lasar may hear the earth, but perhaps the message is really a personal one, just for him.

What then of conversation with the earth? In whose province does it lie? Can it be reconciled with science—or is it science? Is science the arrow that approaches the reality of the universe—or one vehicle among many? Are Zen archers and true dowsers bearers of unique gifts, or do they point a way to more immediate and satisfying understanding that may be employed by many? Perhaps it doesn't matter. We must each develop our own dream of the earth and find a way to it—mystically, empirically, through revelation or evolution—and share our findings with each other. If we don't, then it may be that neither we nor the earth will speak anymore.

Approaches to the Universe

It must be admitted, because it is obvious, that the "geological dream" has its own definite bias. When a geologist as a geologist looks at the world about him, what he sees are patterns and processes. The patterns exist in the shapes, dimensions, colors and textures of rocks and landscape; the processes are manifest in the swirl of water and wind, the sudden collapsing of hillsides, the crackling flow of ice, the eruption of volcanoes, the gradual pervasive action of chemical decay. What the geologist then tries to do is bridge the gap between pattern and process–to explain one in terms of the other; he tries to understand what the patterns imply as to conditions of formation and historical sequence, and also on occasion, he reverses the procedure and attempts to predict future processes that may evolve from existing patterns. The methods he employs are diverse, but they all operate under the assumption that a link between observed patterns and processes is possible, and that these patterns and processes may *only* be explained in terms of each other. This *modus operandi* has been singularly successful in producing a coherent and internally logical view of the earth. I have referred to it before as the Law or Doctrine of Uniformitarianism, the touchstone of modern scien-

Arches National Monument, Utah

tific geology. But which is it, Law or Doctrine? And is it the only approach to earth phenomena? And has it always been a help and never a hindrance? Most geologists would unhesitatingly answer the last two questions in the affirmative, while at the same time readily conceding that Uniformitarianism is an arbitrary assumption, not subject to proof. Thus, the answer to the first question, whether Uniformitarianism is Law or Doctrine, is given: it is Doctrine, if by that is meant *dogma*, for it is clearly accepted on faith—because it is convenient. That is, when applied it produces results of a kind that are considered desirable.

How does Uniformitarianism work? Consider its application to the problem of whether or not a particular river has cut its own valley. Imagine a river flowing in a deep gorge on either side of which a sequence of layered rocks is exposed. Furthermore, imagine that the sequence and character of the layers is the same on each side of the valley. Such a situation may be explained in a number of ways. First an assumption must be made as to whether or not the layers on each side of the valley were originally connected. If the assumption is that they were, then it may be claimed that the river, flowing over a great number of years, gradually, through its powers of erosion, removed the material missing from the space now occupied by the gorge. On the other hand, starting with the same assumption as to the original continuity of the layers, it is quite possible that the layers were separated and the gorge created by some natural process that *pulled* the land apart, and that the river only occupied the gorge after it was formed.

A second assumption might be that the layers on each side of the valley were never connected; that is, that the processes that created the layers on each side of the gorge never operated at the actual site of the gorge itself. The gorge, therefore, resulted from an *absence* of process.

A third possibility is that the river, the gorge and the layering were produced supernaturally in a manner not comprehensible to man.

None of these alternative explanations of the relationship between the river and its valley are subject to "proof"; which one is accepted depends upon what you are after. If the river and its gorge are the result of supernatural activity, then the most logical approach to knowledge of rivers and gorges is not to study *them*, but to study the supernatural. If the river and the gorge resulted from natural processes, then evidence may be sought at the site, at other similar sites, or even experimentally in a laboratory. That is, Uniformitarianism may be invoked.

If the gorge was gouged out of once continuous layers by the action of flowing water, then measurements of rates of water erosion and estimates of the length of time available for such a process to operate should jibe with this proposed theory of origin—and in many instances they do. They do not prove the theory, but they suggest its feasibility.

If the gorge was created by a pulling apart of once continuous layers, then forces capable of performing such a feat must be sought, and auxiliary evidence examined. There are places where rivers seem to have originated in this manner. The River Jordan, for instance, flows in what seems to be a pulled apart crack in the earth's surface.

The possibility that the gorge is the result of non-deposition of layers, layers that were deposited contiguously on either side of the gorge but not within the gorge, is unlikely in most environments, but is conceivable if the gorge lies in what is or once was glacial terrain: an extraordinarily elongated block of ice might have been left in an area subsequently the site of brief, rapid sedimentation. When sedimentation ceased and the block of ice melted, a gorge may have been left. Again, evidence for such an origin would have to be carefully examined.

The last three natural explanations all stem from a Uniformitarian approach: they attempt to relate pattern to observable process. Knowledge of which particular explanation is correct depends upon availability of evidence and correct interpretation of that evidence. It is quite possible that no evidence is available, or that conflicting evidence or misleading evidence is all that is available, and that all conclusions must be tentative and perhaps wrong. Uniformitarianism provides a way of looking at things but does not guarantee unique solutions.

Patterns presumably formed by surface or near-surface, observable processes may be simply explained through Uniformitarianism, but rock patterns and earth phenomena generated by processes taking place beyond the sphere of human observation present a more difficult problem. How can Uniformitarianism be applied to activities proceeding deep within the earth's interior? How can Uniformitarianism be applied to events occurring in the remotest past—when conditions on the planet or in the universe may have been quite different from what they are today?

In the consideration of problems where direct observation is impossible and indirect observation difficult, Uniformitarianism is not abandoned, but rather becomes the starting point for long chains of reasoning. It becomes a statement of the observer's belief in the validity of natural explanations, in the permanence of natural laws, and in the reasonableness of scientific logic and of the world.

When the first two astronauts reached the moon one of them picked up a piece of moon rock and described it as containing *vesicles;* by that he meant it was full of small holes similar to that found in earthly lavas. But in the new, alien environment in which he found himself, what right did the astronaut have to assume the validity of earthly processes? Shortly after, it was suggested that perhaps these holes in the moon rock were the result of micro-meteorite penetrations, a mechanism that does not affect rocks on our own planet. Subsequently, the first explanation was returned to: that the holes were vesicles in a moon lava. But aren't other explanations possible? After all, surely we don't yet know what are or have been in the past "normal" moon-surface processes. The astronaut's initial statement was quite legitimate based upon his own past experience and was also perhaps a way of making a strange environment somewhat less strange. His experi-

McCabe Lakes, Yosemite

ence (and all of humanity's experience) is of necessity limited. Furthermore, it is an entirely human tendency when confronted with the new to interpret it in terms of the familiar; that is, to apply or misapply the concept of Uniformitarianism. After all, what other equally fruitful approach is there for us?

To some people, approaches that guarantee unique solutions are considered more fruitful. Fundamentalist preachers, for example, often pour scorn upon the confusion of science. To scientists, the fact that complex questions, such as the origin of life, invoke many tentative, often seemingly contradictory theories is hardly surprising; it is the way in which science works. From many pulpits, however, this lack of unanimity is seized upon as weakness, evidence for the invalidity of science and of Uniformitarianism. Strangely enough, to many people it is a revelation to learn of controversy within the scientific community. It is only too common to hear laymen declare that "science says," as if science had one clear voice. When they hear of argument within scientific ranks it is as if an Achilles' heel had been exposed, and the whole edifice of "science" in their eyes comes tumbling down—to be replaced, no doubt, by the secure knowledge found within the Holy Book. Science has many Achilles' heels, but more likely they are related to dogmatism than to doubt.

Religious fundamentalism is not the only source of unique solutions. There are always those willing to offer imaginative, grand schemes for answering unanswered questions. The type of logic often employed in such schemes is illustrated succinctly by Father Sogol (*logos* spelled backwards) in René Daumal's beautiful allegorical fragment *Mount Analogue*. In this novel, the narrator has deduced that according to mythic tradition "the Mountain is the bond between Earth and Sky. Its solitary summit reaches the sphere of eternity, and its base spreads out in manifold foothills into the world of mortals. It is the way by which man can raise himself to the divine, and by which the divine reveals itself to man." However, the narrator continues, in order that a mountain be the ultimate symbolic mountain (to be called Mount Analogue) and retain its significance, it must remain inaccessible "to ordinary human approaches." Mountains such as Sinai, Olympus and Everest, once occupied by divinity, lost their peculiar essence and potential as they were conquered by man. The ultimate symbolic mountain that unites heaven and earth must have a summit that is inaccessible. But it must also have a base that is accessible to ordinary human beings. If its base were not accessible, the climb towards heaven could not be attempted, and therefore the whole mountain could have no relevance to mere mortals.

At this point Father Sogol takes over. Since humans must strive towards heaven, the way—that is, the mountain—must exist and be capable of being found. The method to employ in finding it, Father Sogol declares, "consists in regarding the problem as solved and deducing from the solution all logical consequences." Since no mountain having the characteristics of Mount Analogue is described in any geographical journal, and since for metaphysical reasons the mountain must exist, it must, therefore, be invisible. The mechanism of its invisibility and the coordinates of its location are then deduced in a quite logical and successful manner, and its approach undertaken.

Uniformitarianism here has flown out the window—or has it? Doesn't pattern still suggest process and process explain pattern? They do, but the patterns include not only objective reality but also subjective reality—the needs and aspirations of the human spirit—and the processes involve local warping and suspension of normal physical laws. It is in this that Uniformitarianism is abandoned.

René Daumal's *Mount Analogue* is described by its author as fiction, but there are others who employ similar species of logic to explain earth phenomena who are quite in earnest. Charles Hapgood, for instance, invokes an earth whose crust ever so often slides ninety degrees over the underlying mantle; Immanuel Velikovsky looks outward to earth-comet collisions to explain the world he sees about him. Each of these gentlemen attempts to answer certain legitimate questions: Hapgood is concerned with

(Overleaf) Goosenecks of the San Juan River, Utah

the occurrence of Ice Ages and problems of paleoclimatology; Velikovsky wishes to explain similarities among primitive myths of diverse, widely separated peoples, and in doing so, creates his own peculiar geology. In each case the solutions offered are comprehensive, carefully reasoned, and have an elegant unity of thought and presentation. All evidence points to the conclusion, each step of the argument fits into the next with machined precision.

It is not in their exoticism that these theories depart from Uniformitarianism, for they are no more bizarre, ultimately, than the theory of continental drift currently in vogue in respectable geologic circles; rather it is in the nature of the evidence they present and their order of priorities. In scientific as opposed to pseudoscientific polemic, evidence has priority over hypothesis. That is, hypothesis must conform to evidence, not evidence to hypothesis. If evidence contrary to a given hypothesis is found, the hypothesis must (or should) be altered or rejected. This does not mean that when contrary evidence is found the hypothesis is immediately rejected; it may be retained tentatively in the hope that further evidence may eventually explain the "contrary" evidence in terms of the theory. Indeed, many hypotheses are arrived at intuitively, through a "flash of insight," and may be maintained in the mind of the investigator in spite of much initial "contrary" evidence, and later be vindicated—the human mind often follows a devious route. But a legitimate researcher will not consciously suppress contrary evidence. Certainly he will not publish a hypothesis until he has satisfied himself that all the available evidence is in accord with it. Indeed, if he is properly conscientious, he will search for contradictory evidence to test the validity of his ideas. Hypothesis may precede evidence in time but not in importance.

Earth theories of the Hapgood-Velikovsky type, however, are based upon evidence carefully selected to support a theory. Contrary evidence is suppressed as unimportant or irrelevant. But could their ideas be true? Of course they could. But whose, Hapgood's, Velikovsky's or someone else's? By what criteria can they be judged, since evidence is treated lightly, Uniformitarianism is rejected and degree of truth depends largely upon depth of inspiration? If one-world view is accepted, others must be rejected, and decisions as to which must be made almost upon an aesthetic basis.

It is perhaps significant that neither Hapgood nor Velikovsky were trained as geologists, yet their works have had considerable distribution among a lay public who might take them for geologists. Furthermore, the facility of their styles and the apparent (but misleading) detailed documentation of their ideas makes it particularly difficult for the non-expert to see that what they have written is pseudo science rather than science, that when looked at critically their work falls apart at every seam. Interestingly enough, Albert Einstein was taken with the ideas of both these men, which suggests that even genius must be suspect outside its own field.

Avenues of scientific approach that follow the mundane dictates of Uniformitarianism may, as previously hinted, lead to conclusions far from mundane. In 1924 Alfred Wegener, a meteorologist, after considering the similarity in outline between the east coast of South America and the west coast of Africa, together with much other evidence, proposed that the two continents were once joined together and subsequently drifted apart. The concept of "continental drift," after an initial stir of interest, was largely rejected throughout the geologic world because no direct evidence for and no process capable of generating the drift could be found. Uniformitarianism required that the idea be put on the shelf. Over the years, however, more and more evidence accumulated that seemed understandable only in terms of continental drift. But the scientific community is very demanding, and even though numerous theoretical mechanisms capable of causing drift were proposed, the idea was still not generally accepted. It was not until the advent of earth-circling man-made satellites that geologists as a whole were finally convinced. In successive years advanced surveying techniques employing satellites were used to measure distances between continents, and it was found that the distances actually changed with time—in a manner consistent with the idea of continental drift: the demands of Uniformitarianism had finally been met. It might be argued that if a less rigorous, non-Uniformitarian Hapgood-Velikovsky approach had been employed, the state of the science would have advanced more rapidly. But that, of course, is an argument based upon hindsight; continental drift might have turned out to be a false idea, and the science would have been headed up a blind alley.

The Hapgoods and the Velikovskys really have their place in the literature of fantasy. It is unfortunate that they claim to be scientific and that an uncritical public often accepts them as such. Choosing a Uniformitarian approach is not opting for dullness. An earth with churning insides and wandering continents, a moon with pock-marked pebbles and gigantic craters, landscapes with deep gorges and thundering streams are not lacking in the fantastic, nor is a geology rife with dissent showing its Achilles' heel. The Uniformitarian approach has no claim to unique truth or even to any truth, but it is certainly an interesting and fruitful way to view the universe.

The Major Prize

It is probably crass to ask what the principal rewards are in being a geologist—what is the major prize—but the question is a legitimate one. At some level every person must consider the connection between himself and what he is doing—the life he leads—and consciously or instinctively arrive at a decision as to whether or not he is misplaced, whether the chance of reward is there. The conclusion he reaches is not abstract; it is clearly revealed in his own sense of happiness or unhappiness, serenity or desperation.

The major reward in geology, as I have found it, is the deep intimacy that can arise between a man and his natural surroundings, an intimacy that stems from both detailed physical contact and continued theoretical and philosophic contemplation. The geologist's job is to know the land; hopefully he may come to love it too. If he can love it, for what it is and the way it is without regard for what it may yield or produce, what higher reward could there be? It is a reward that lets a man be at home wherever he is—and sets him free.

The nature of geologic field work and the character of the materials dealt with are what give geology this potential. Un-

like many other disciplines, there is opportunity for intimate contact with substance as well as abstraction, the possibility of realization that man and the earth are indistinguishable and inseparable, each parts of a whole not often recognized. In such approaches to reality resides geology's liberating quality.

What the geologist seeks is knowledge of the way the land is put together: how it is now, how it was in the past, how it is evolving towards the future. It is a process of endless initiation into the paramount meaning and control of rock.

As a geologist gains experience–that is, as he becomes familiar with particular materials and their local structural and erosional styles–he learns to infer which rock formations underlie specific areas without having to see them. He steeps himself in the character of a region until the presence or absence of a certain type of vegetation, the color of the soil, the angles of the horizon against the sky, the spot where a spring emerges from the ground, the location of a river bed, the acreage a farmer chooses to plow, the extent of mosquito-ridden swamps –all begin to tell him a tale of rock composition, rock texture and the distribution of rock units in space.

In the vicinity of Saugerties, New York, for instance, there is a battleship-gray limestone called the New Scotland which has three siliceous horizons, each full of roughly egg-shaped cherty nodules. Wherever the surface of the land is at an angle to and cuts across the New Scotland, the three cherty layers stand up as three subtle, but continuous, parallel ridges, masked somewhat by grass and trees, but obviously there–when you learn to look for them. In the same region, but further south, the Wilbur-Rosendale and the Rondout formations are limestones that contain the correct admixture of clay for natural cement and were quarried and mined extensively as such during the last century. As a result, wherever these two formations appear at the surface, their presence is manifest by deep, man-made pits and shafts that appear along the roads and in amongst the rotting debris of the forests. Stagnant sunlit pools and flowery swamps imply

Junction of the Rio Grande and the Rio Frijoles, New Mexico

the subterranean presence of the High Falls shale. Nearby, the Shawangunk conglomerate, a thick formation composed of quartz pebbles cemented together by a quartz cement, forms the dominant ridges of the area.

Thus all the details of the local topography echo mineral and chemical variations within the underlying rock. Every minute change in the proportions of silica, magnesia, lime or clay affects each formation's ability to withstand the ravages of the elements. As the geologist walks out and follows the land, all this is impressed indelibly on his mind, muscles and lungs. An intense satisfaction comes with learning the logic of each bump and ripple in the earth's surface, of being able to successfully predict what will lie on the *other* side of each hill or the far slope of a broad valley. When he is wrong the struggle to understand why leads to greater accuracy in the future. Gradually the land becomes his, revealing its secrets, displaying the art and cunning of all its hidden mechanism.

In soft green and wooded areas, such as parts of New York State, geology is redolent of mystery and decay, and the geologic art is a difficult one. Control of rock type and structure over topography is sure, but boundaries may be vague rather than sharp. Incisions of erosion are healed by forests and grass; the inexorable creep of soil and loose rock quickly smooth off acute angles and shape the land into a round and gentle place. The eye has to be sharp, the glance studied.

But in arid regions where the land is bare, the control of rock is obvious and spectacular. Here there is no mask. Each irregularity, each strength and weakness stands open to view. The chain of cause and effect is measurably shortened. The relationship between structure and form is revealed—or if it is not revealed, that fact is also clearly evident. Hard rocks make no pretense at modesty; they jut out as great spires and pillars or form the tops and walls of precipitous cliffs. Soft materials disintegrate quickly to intricately eroded slopes and deep crevasses. Without doubt the rocks are lovely and emerge in full glory.

Bridge Canyon, Utah

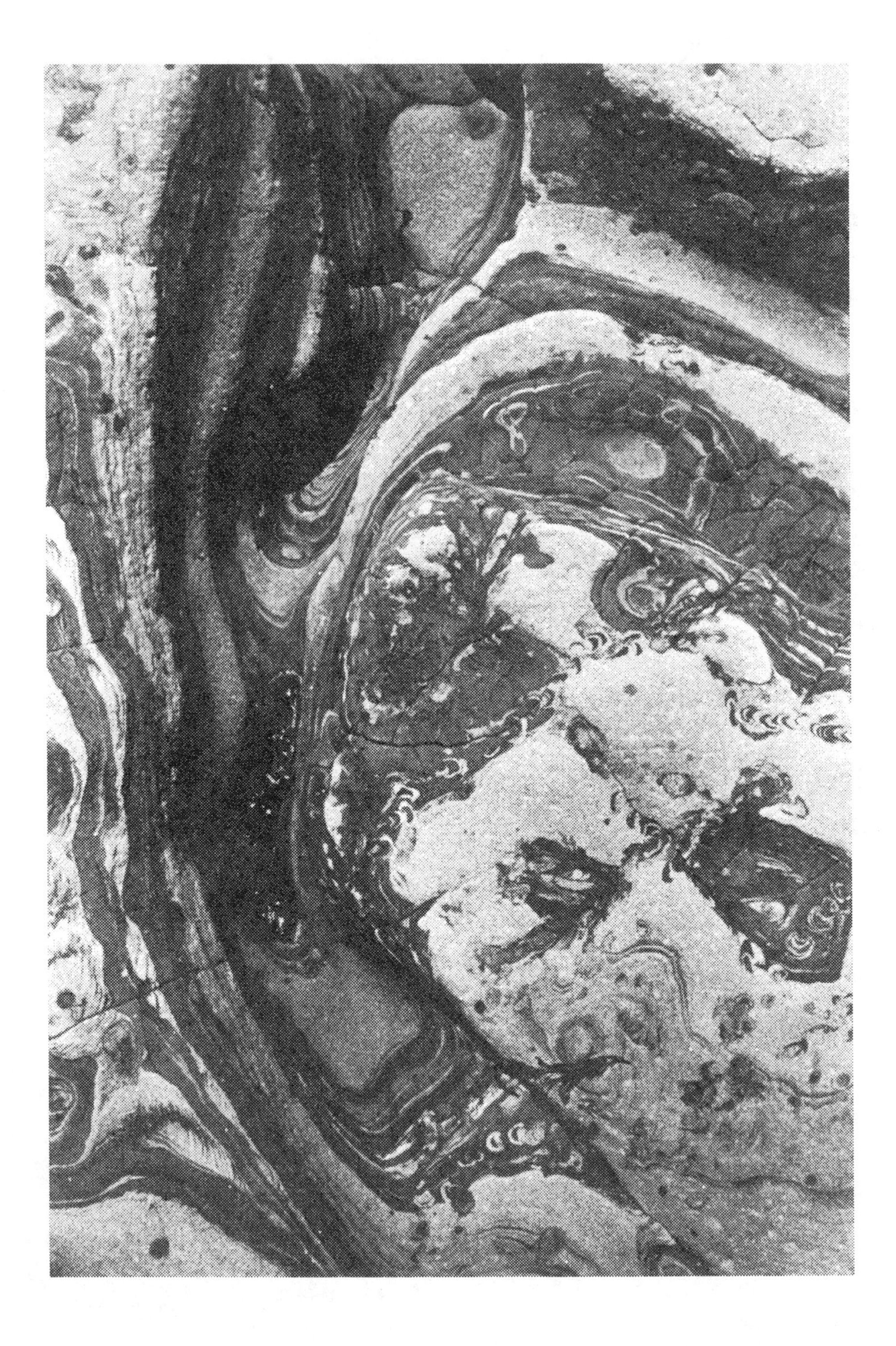

The geologist at one place may find himself wandering through a landscape where every rise and ridge is crowned by dark towering spines, or out on a broad flat plain divided into innumerable segments by a network of black, finlike slabs. A blow from his hammer reveals these natural fences as the surface expression of vertical basalt dikes.

Elsewhere, he may suddenly emerge at the rim of a plateau and look out over a panorama that climbs and falls to the horizon in a series of giant steps, each capped and supported by the resistant remains of ancient lava flows. Where resistant rocks —basalts, limestones or sandstones—are tilted somewhat, they form linear or curved ridges that stretch, almost impassable, into the distance.

Unusually soft rocks, such as poorly consolidated clays, sands and volcanic ash, become deeply and intricately dissected by the multitudes of streams that flow briefly after desert storms, and develop into "badlands": stark declarations of labyrinthan folds or uncrossable mazes of vertical cones that taper steeply upward to sharp points.

Thus, in arid areas the force of rock cannot be denied; it is ever-present and overwhelming. Each rock type has its characteristic expression. Its role is clear. For the geologist the girders of the earth emerge naked, the function of each, in the structure and integration of the landscape, manifestly visible.

Just as differential resistance to weathering and erosion on a large scale determines and shapes the landscape, the same processes operate on the smallest scale and produce the minute details that characterize the texture of rocks and lend them their charm and individuality. In limestone regions it is not uncommon to find rock surfaces decorated with an intricate fretwork of lacelike septa that rises several millimeters above the surrounding matrix. The septa are the surface expression of quartz veins that riddle the limestone; being less soluble than the calcite surrounding them, they stand out in relief as the material adjacent to them is removed.

In north-central New Mexico, there is a sandstone layer whose

Pebbly Beach, Point Lobos, California

surface, with uncanny detail, reproduces in miniature the spectacularly eroded high plateaus of neighboring Utah and Colorado. On a scale measured in inches, there are natural bridges, caves, rock windows, towering spires, benches, overhanging cliffs, deep crevices, hidden basins and tortuously twisted valleys. In boulder-size pieces of rock, whole worlds may be carved. The cause? A cement of variable solubility, which in places holds the sand grains together, in others lets them wash away.

Sediments decorated by concretions that stick out like marbles, tennis balls or the backs of turtles are not uncommon. If broken open, many have hollow interiors—in some cases filled with clay, in others lined with shining crystals.

Weathering reveals textures in igneous and metamorphic rocks just as curious as those in sedimentary formations. Staten Island in the city of New York consists in part of a greenish metamorphic rock called serpentine. In many places the serpentine has been highly fractured, and the fractures filled with a network of criss-crossing quartz veins. At one locality the serpentine, particularly vulnerable to decay, has been more or less completely removed. All that has been left is the quartz which, bereft of its matrix, stands as a cavernous, skeletal framework.

Igneous rocks known as porphyries often show surfaces studded with crystals. Porphyries are rocks that have two distinct grain-sizes: large grains or crystals called phenocrysts lie in a finer-grained background. If the background is less resistant to weathering than are the phenocrysts, the latter will jut out like jewels on a shield. If the phenocrysts are the less-resistant part of the rock, a pitted rotten appearance will develop as they are dissolved.

In the Grand Tetons of Wyoming, rocks with hundreds of "eyes" composed of single, black magnetite crystals embedded within oval masses of white feldspar stare unprovoked at all who pass.

How can the geologist fail to be continually delighted by the shapes and surfaces he comes across? His profession demands that he be aware of these endless variations, and at some level he must respond to them, emotionally as well as intellectually. If he doesn't, his work, as a reflection of his inner self, must inevitably suffer.

In the past, the earth, rocks and wilderness were often as not considered The Enemy. In the nineteenth century D. D. Dana predicted that ". . . Man of the future is Man triumphant over dying Nature, exulting in the freedom and privileges of spiritual life. . . ." Modern man is learning gradually and painfully, that if nature dies, he dies too–spiritually and physically. There is little that man can do to improve the earth. At best he can harmonize with it; at worst he can desecrate it. Fundamental to continued and improved existence is the care and conservation of our environment, a reality that can only successfully stem from contact and love–of life, of the earth.

Geology provides the opportunity, the necessity, indeed the obligation of showing the way. As the geologist geologizes, the manifestations of rock–whether textural details that his eye can see and his thumb feel in a hand specimen, or large-scale labyrinths through which he hikes and climbs–all gradually become part of him, intuitive knowledge that lets him know where he is and what he may expect. They form the basis for the intimacy that springs up between him and the earth–the intimacy that is, after all, his and, ultimately, humanity's major prize.

References

F. D. ADAMS, *The Birth and Development of the Geological Sciences* (1938). Reprint based on 1st ed. (New York: Dover Publications, 1954).

W. B. N. BERRY, *Growth of a Prehistoric Time Scale Based on Organic Evolution* (San Francisco and London, W. H. Freeman and Co., 1968).

HARRISON BROWN, *The Challenge of Man's Future* (New York: Viking Press and Macmillan Co., of Canada, Ltd., 1954).

P. E. CLOUD, JR., "Atmospheric and Hydrospheric Evolution on the Primitive Earth," *Science*, 160 (1968), pp. 729–36.

BARRY COMMONER, *Science and Survival* (New York: Viking Press, 1963).

D. D. DANA, *The Geological Story Briefly Told*. An introduction to geology for the general reader and for beginners in the science (New York and Chicago: Ivison, Blakeman, Taylor and Co., 1879).

CHARLES DARWIN, "On the origin of species by means of natural selection or the preservation of favoured races in the struggle for life" (London: John Murray, 1859).

RENÉ DAUMAL, *Mount Analogue*, translation and introduction by

Roger Shattuck, first published by Vincent Stuart Publishers, Ltd., 1959 (San Francisco: City Lights Books, 1968).

CHARLES H. HAPGOOD, *Earth's Shifting Crust.* A key to some basic problems of earth science (New York: Pantheon Books, 1958).

J. M. HARRISON, "Nature and Significance of Geological Maps," *The Fabric of Geology*, C. C. Albritton, Jr., ed. Prepared under the direction of the Geological Society of America in commemoration of the Society's 75th Anniversary (Reading, Massachusetts, Palo Alto, London: Addison-Wesley Publishing Co., Inc., 1963).

EUGEN HERRIGEL, *Zen in the Art of Archery*, trans. by R. F. C. Hull (New York: Pantheon Books, 1953).

D. L. LAMAR AND P. M. MERIFIELD, "Cambrian Fossils and Origin of Earth-Moon System," *Bulletin of the Geological Society of America*, No. 78 (1967), pp. 1359–68.

TED LASAR, *Life Was Not Always a Tragedy* (1963).

C. R. LONGWELL, R. F. FLINT, AND J. E. SANDERS, *Physical Geology* (New York, London, Sydney: John Wiley and Sons, Inc., 1969).

THOMAS MANN, *Joseph and his Brothers*, trans. by H. T. Lowe-Porter (New York: Alfred A. Knopf, 1934).

BRIAN MASON, "Meteorites," *American Scientist* 55 (1967), pp. 429–55.

JOHN PLAYFAIR, *Illustration of the Huttonian Theory of the Earth* (1802). A facsimile reprint with an introduction by George W. White (New York: Dover Publications, Inc., 1956).

OSWALD SPENGLER, *Decline of the West.* Abridged edition by Helmut Werner. English abridged edition by Arthur Helps from translation by Charles Francis Atkinson (New York: Alfred A. Knopf, 1962).

IMMANUEL VELIKOVSKY, *Worlds in Collision* (New York: Dell Publishing Co., by arrangement with Doubleday & Co. Inc., 1950).

——— *Ages in Chaos* (New York: Doubleday & Co. Inc., 1952).

——— *Earth in Upheaval* (New York: Doubleday & Co. Inc., 1955).

G. J. WASSERBURG, H. G. SANZ, AND A. E. BENCE, "Potassium-feldspar phenocrysts in the surface of Colomera, an iron meteorite," *Science*, 161 (1968), pp. 684–87.

ALFRED WEGENER, *The Origin of Continents and Oceans*, trans. by John Biram (New York: Dover Publications, 1966).

For recent discussions of continental drift, see H. Takeuchi, S. Uyeda, and H. Kanamori, "Debate about the earth, approach to geophysics through analysis of continental drift," trans. by Keiko Kanamori, based on *Science of the Earth* by Takeuchi and Uyeda (San Francisco: Freeman, Cooper and Co. and Nippon Hoso Publishing Co., rev. ed., 1970), and D. H. and M. P. Tarling, *Continental Drift* (New York: Doubleday & Co. Inc., 1971).

Index